MW00575216

THE LIVES OF

MOTHS

THE LIVES OF MOTHS

A NATURAL HISTORY OF OUR PLANET'S MOTH LIFE

Andrei Sourakov
& Rachel Warren Chadd

PRINCETON UNIVERSITY PRESS
PRINCETON AND OXFORD

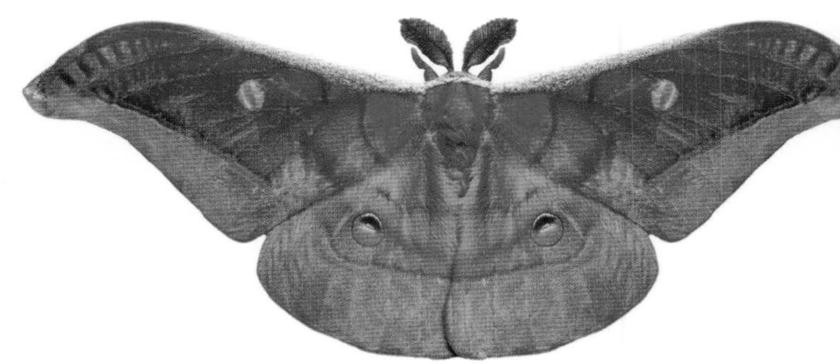

Published by Princeton University Press
41 William Street, Princeton, New Jersey 08540
99 Banbury Road, Oxford OX2 6JX
press.princeton.edu

Library of Congress Control Number 2021948357
ISBN 978-0-691-22856-3
Ebook ISBN 978-0-691-23036-8

Typeset in Bembo and Futura

Printed and bound in China
10 9 8 7 6 5 4 3 2 1

British Library Cataloging-in-Publication Data is available

This book was conceived, designed, and produced by
UniPress Books Limited
Publisher: Nigel Browning
Commissioning editor: Kate Shanahan
Project manager: Caroline Earle
Designer & art directon: Wayne Blades
Picture researcher: Sharon D'Ortenzio
Illustrator: John Woodcock
Maps: Les Hunt

Cover photos:
Igor Siwanowicz
(front cover: *Actias dubernardi*);
Shutterstock /Matee Nuserm
(back cover and spine: *Mangina argus*).

CONTENTS

INTRODUCTION

The world of moths

It's the 1970s, and I am walking from school past a high-rise apartment building. I suddenly stop in my tracks, while my heart begins to race. On the brick wall, I detect the unmistakable triangular shape of the Red Underwing moth. Slowly, not to startle it, I approach; even more slowly, I extend my hand and touch its hairy back. The moth flicks up its forewings, exposing a flash of red from hind wings normally hidden from view, in a desperate attempt to scare me off. When I touch it again, the moth zooms up and perches high above the ground, instantly becoming just another "scar" on the bark of the tall poplar.

The diet of their caterpillars sustains the intimate connection between moths and plants. Which plants occur where is determined by numerous factors, from geography and evolutionary history to soil composition and levels of sunlight and water. And while different continents may have different moth faunas, each moth community—whether in a rainforest or desert—bears a distinct imprint of its habitat. According to both habitat and geographic region, moths also interact with a host of other organisms—as large as grizzly bears and as tiny as viruses.

In the present volume, we first examine the moth's four stages of development, from egg to adult, and its biology and behavior in different environments, before venturing to explore examples of moths found in vast habitats of tropical forest, grasslands, deserts, and tundra. Certain moths have undergone interesting adaptations to occupy aquatic habitats, and it may come as a surprise to many that some species develop in water. There are also moths that live in sloths' fur, drink bird tears, or even, as caterpillars, predate on wasps or mollusks. The secret world of moths is truly remarkable!

Of course, moths are mobile creatures, and many of them move between habitats in search of nectar for themselves or plants to lay their eggs on. Some species even migrate seasonally and others are, like us, highly versatile, and have formed different races specifically adapted to the habitats of their geographic region. These, however, are exceptions rather than the rule, and I hope that showcasing moths as integral parts of their respective ecosystems will help in appreciating these species' roles in their environment. Today, when natural habitats are disappearing at an unprecedented rate, yielding to those created by humans, underscoring the connection between habitat type and the unique species that they harbor becomes vitally important. Only by conserving habitats can we preserve the precious species that inhabit them.

Andrei Sourakov

↑ → Two beautiful moths that the author first encountered as a child inside the city: the Red Underwing (*Catocala nupta*) that develops on poplar (top) and the Elephant Hawk Moth (*Deilephila elpenor*), whose caterpillars eat rosebay willowherb along rail tracks and in urban wasteland.

What is a moth?

The evolution of moths—insects of ancient lineage in the order Lepidoptera—is intimately entwined with that of plants. While their diversification occurred during the rise of flowering plants from around 130 million years ago, gymnosperm plants 70 million years earlier appear to have played an important role in their origins and speciation.

THE ORIGINS

It was the recent discovery of a 200 million-year-old fossilized moth in Germany that pushed back the probable date of Lepidoptera origins and prompted the hypothesis that during the Jurassic period, before there were flowers, moths developed a sucking proboscis to sip droplets of moisture from the tips of immature seeds of plants related to today's conifers. The proboscis—part of the maxilla (mouthparts) called galeae, zipped together into a straw-like organ—continued to evolve and today distinguishes most (though not all) moths and butterflies from other insects, whose classification has traditionally been based on mouthparts. Some moths have retained their chewing mouthparts, but they are in a minority.

MOTH OR CADDISFLY?

Their closest relatives are Trichoptera (caddisflies), which also developed in the early Jurassic period, and together with Lepidoptera form a group called Amphiesmenoptera. While the two share some characteristics, such as larvae that can produce silk, there are major differences; the wings of moths, for instance, are covered in scales, while those of caddisflies are hairy.

MOTHS VERSUS BUTTERFLIES

People often wonder how butterflies relate to moths and may be surprised to know there are no major differences. Butterflies, which evolved from a common ancestor about 110 million years ago, form a group of just eight families within Lepidoptera, otherwise comprised of some 130 moth families, so are simply an offshoot of the moth evolutionary tree. Based on their genetic analysis, plume moths (Pterophoridae) are probably most closely related to butterflies. Like moths, certain butterflies, including many skippers and the American moth-butterflies (Hedylidae) fly at night, while numerous moths have independently evolved day-flying habits at least 30 times during their evolution.

ECOLOGICAL IMPORTANCE

Being more ancient, moths have experienced and adapted to a far greater range of conditions and environments than butterflies and thus are more diverse in their morphology and lifestyles. And while the caterpillars of a few moth species—those that eat crops—may have given moths a bad name, most species exist in balanced relationships with their ecosystems, playing crucial roles as pollinators and food for vertebrates. Many species have developed such intimate relationships with their hosts and the flowers they pollinate that neither can exist without the other. As this book reveals, across diverse ecosystems, moths play a crucial role.

↑ Among more advanced moths are the bombycoids, such as this Hummingbird Hawk Moth (*Macroglossum stellatarum*) with a fully developed proboscis that is used to sip nectar in flight.

↖ A member of the mandibulate archaic moth family Micropterigidae, this Marsh Marigold Moth (*Micropterix calthella*) as an adult feeds on pollen grains of various plants.

Moth classification

Of the millions of animal species on Earth, two-thirds are insects. After Coleoptera (beetles), Lepidoptera (butterflies and moths) and Hymenoptera (ants, bees, wasps) are the two most numerous orders, and together these three orders are responsible for half of all insect species.

Among Lepidoptera, in terms of species, moths outnumber butterflies by more than eight to one. Taxonomists attempt to group animals so that each category, such as family or genus, is monophyletic (includes all descendants of a single ancestor and nothing else). "Moths" is not a category as such, while butterflies are. Why? Because butterflies (with their seven families) are an offshoot of moths that derived from a single ancestor, branching off moths' evolutionary tree around 100 million years ago.

The approximately 150,000 species of moths are grouped in over 120 families, which in turn are divided into subfamilies and genera. This classification changes constantly with better understanding of the evolutionary history—morphological studies of the past 250 years are now supplemented by DNA analysis. While most of the larger moths, such as Saturniidae (saturniids or giant silk moths) and Sphingidae (sphingids, sphinx moths, or hawk moths) have been described, much work remains to describe the diversity of rapidly vanishing, smaller, tropical moths.

A moth family can be tiny or numerous. For instance, the family Endromidae to which the Kentish Glory (*Endromis versicolora*) belongs, contains only about 30 species, while the family Erebidae (erebids) includes tens of thousands of species belonging to diverse subfamilies such as tiger, lichen, and wasp moths (subfamily Arctiinae), underwing moths and their relatives (Erebinae), and tussock moths (Lymantriinae). Superficially, moths belonging to the same family can look very different from each other and can lead diverse lifestyles, but they are unified by more stable morphological characters, such as wing venation.

A SELECTION OF MOTH FAMILIES

Here we list and illustrate a few representatives of the most speciose families mentioned in the book—a more complete list of families can be found on page 281.

ACROLOPHIDAE
Tubeworm moths (acrolophids)

ANTHELIDAE
Australian lappet moths (anthelids)

BATRACHEDRIDAE
(batrachedrids)

BOMBYCIDAE
Silk moths (bombycoids)

BRAHMAEIDAE
Brahmin moths (brahmaeids)

COLEOPHORIDAE
Casebearer moths (coleophorids)

COSMOPTERIGIDAE
(cosmopterigids)

COSSIDAE
Carpenter moths (cossids)

CRAMBIDAE →
Sky-pointing moths (crambids)

DREPANIDAE
Hook-tip moths and casebearers (drepanids)

ELACHISTIDAE
Grass-miner moths (elachistids)

ENDROMIDAE
(endromids)

EREBIDAE ↓
Tiger, lichen, and wasp moths, underwing moths, tussock moths, owlet moths, woolly bears (erebids)

GEOMETRIDAE ↓
Inchworms, butterfly moths (geometrids)

GRACILLARIIDAE
Leaf blotch miner moths (gracillariids)

LASIOCAMPIDAE ↓
Lappet moths or eggars (lasiocampids)

LIMACODIDAE
Slug moths (limacodids)

MEGALOPYGIDAE ↓
Flannel moths (megalopygids)

MIMALLONIDAE
Sack-bearer moths (mimallonids)

NEPTICULIDAE
Leaf miners (nepticulids)

NOCTUIDAE ↓
Owlet moths (noctuids)

NOLIDAE ↓
Tuft moths (nolids)

OECOPHORIDAE
(oecophorids)

PLUTELLIDAE
(plutellids)

NOTODONTIDAE ↓
Prominent moths (notodontids)

PRODOXIDAE
Yucca moths (prodoxids)

PSYCHIDAE
(psychids)

PTEROLONCHIDAE
(pterolonchids)

PTEROPHORIDAE
Plume moths (pterophorids)

PYRALIDAE
(pyralids)

SATURNIIDAE ↓
Silk moths, oak worm moths, buck moths (saturniids)

SCYTHRIDIDAE
Flower moths (scythridids)

SESIIDAE
Clearwing moths (sesiids)

SPHINGIDAE ↓
Hawk moths (sphingids)

STATHMOPODIDAE
(stathmopodids)

TINEIDAE
Fungus moths (tineids)

TORTRICIDAE ↓
Carpenter moths (tortricids)

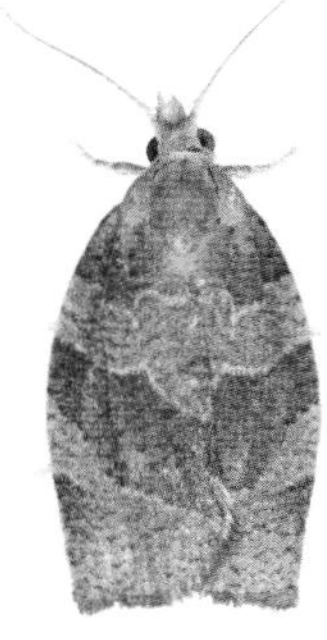

URANIIDAE
Sunset moths (uraniids)

YPONOMEUTIDAE
Ermine moths (yponomeutids)

ZYGAENIDAE ↓
(zygaenids)

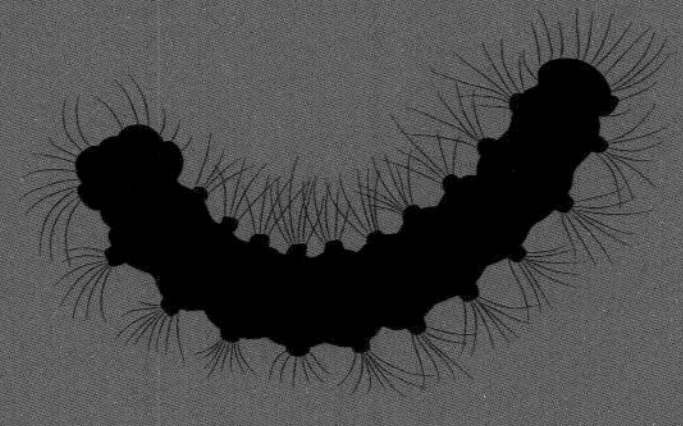

LIFE CYCLE

Eggs and oviposition

Moths may lay eggs (oviposition) singly or in batches of thousands, either glued to the host plant, inserted into it, or dropped from the air nearby by the adult female. Egg shapes can vary from a perfect sphere to something resembling a pancake, flying saucer, or football. They can be white, translucent, or colorful, but need to be inconspicuous, or have some way of repelling predators. The eggs must also withstand environmental pressures, from freezing cold and rain to intense heat, while protecting the delicate embryo of a future moth, often within the confines of a space a fraction of a millimeter in size.

INGENIOUSLY CONSTRUCTED

While we may marvel at the Hagia Sophia and its unique dome and arches as examples of unsurpassed ancient architecture, moths have been masters of such construction techniques for more than 200 million years. When follicle cells within the female's ovaries produce the eggs, the proteins of the future shell are arranged into flattened spiral shapes called helicoids. As a result, the eggs have strong but light "struts and columns" of outer layer (exochorion) and airy structures of inner endochorion, which together form the shell.

Like birds' eggs, the moth egg contains a single egg cell that is fertilized just before laying by sperm which the female stores separately from the forming eggs in a special sac called a bursa copulatrix. Within the eggshell, the embryo is surrounded and protected by a membrane and a layer of wax. Similar to fish eggs, moth eggs have openings called micropyles through which sperm enters during fertilization.

← Females of the subfamily Plusiinae (Noctuidae) may lay hundreds of eggs in one batch—here on sweet clover.

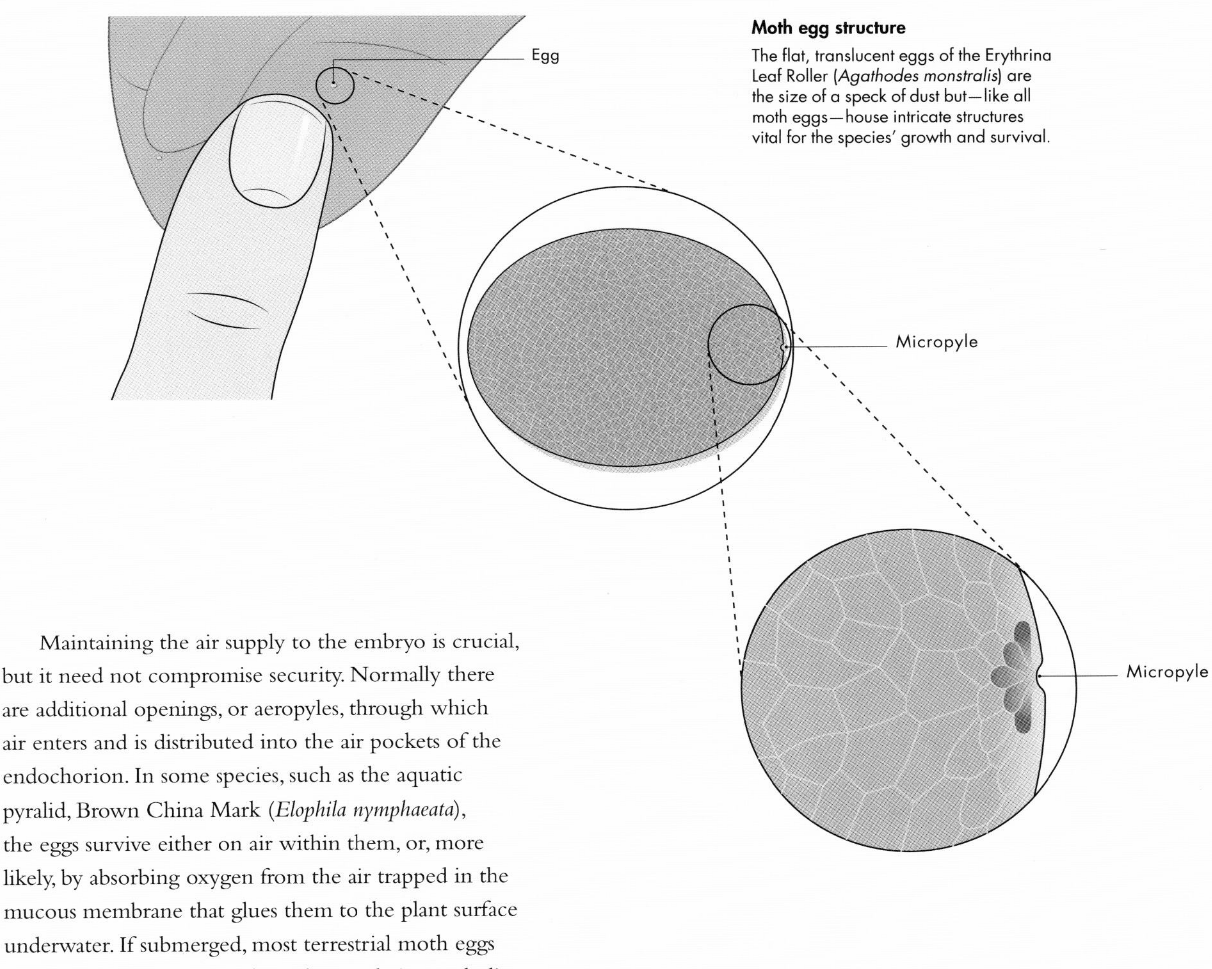

Moth egg structure

The flat, translucent eggs of the Erythrina Leaf Roller (*Agathodes monstralis*) are the size of a speck of dust but—like all moth eggs—house intricate structures vital for the species' growth and survival.

Maintaining the air supply to the embryo is crucial, but it need not compromise security. Normally there are additional openings, or aeropyles, through which air enters and is distributed into the air pockets of the endochorion. In some species, such as the aquatic pyralid, Brown China Mark (*Elophila nymphaeata*), the eggs survive either on air within them, or, more likely, by absorbing oxygen from the air trapped in the mucous membrane that glues them to the plant surface underwater. If submerged, most terrestrial moth eggs can survive for some time by reducing their metabolic rate, living on the air supply stored in their air chambers, or on air trapped against their sculptured surfaces.

MOTH EGG SHAPES

Moth egg shapes vary from spherical to flat. Most are laid on a plant's surface, but some are inserted into plant tissue.

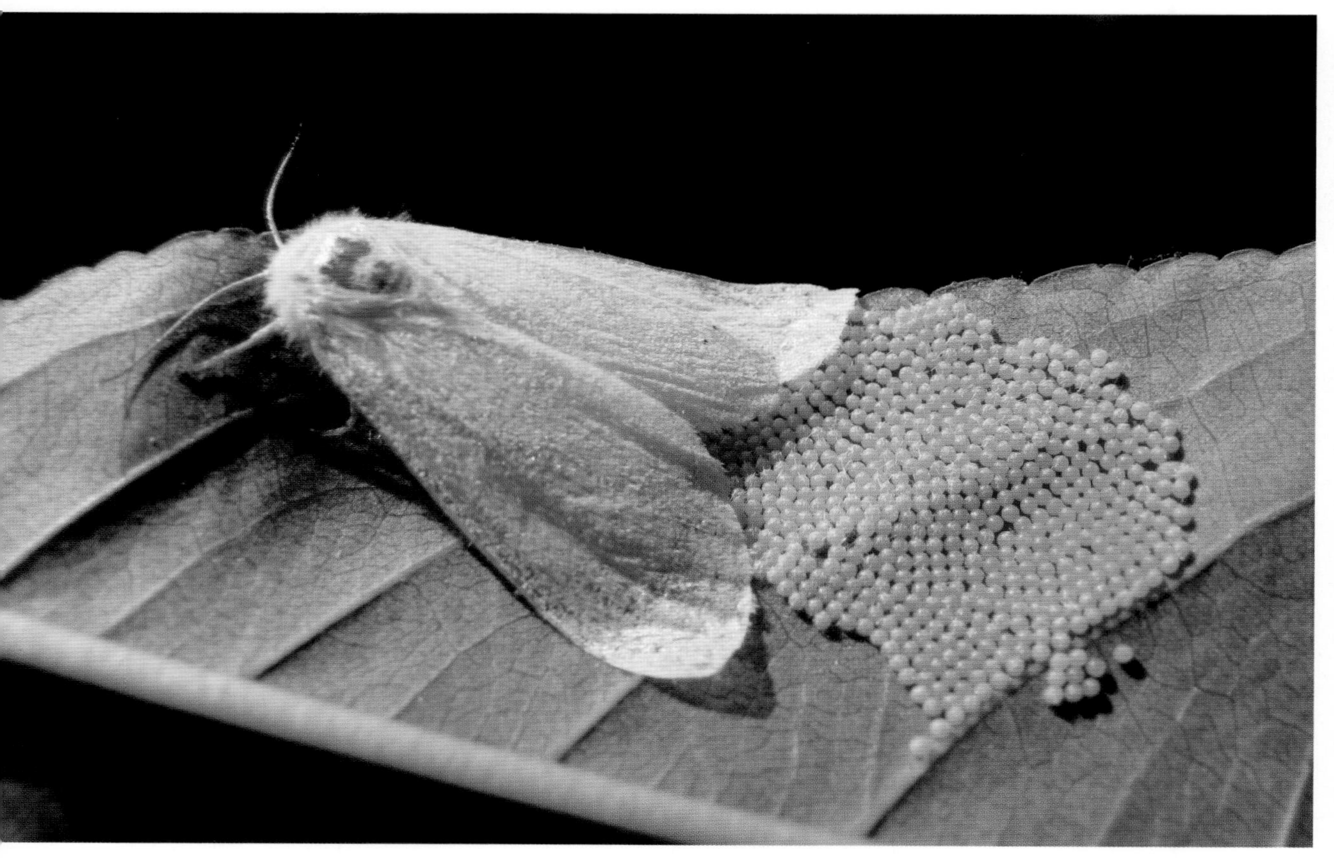

JUDICIOUS LAYING

Sometimes moth eggs are laid singly by a female aiming to distribute them widely among as many host plants as possible over a broad geographic area. This strategy spreads the risk of predation and other dangers, and provides more food for each offspring. Such behavior is more typical of butterflies, whose females fly by day and seek out host plants far and wide. Large, fast-flying hawk moths behave similarly—some species flying by day or at dusk, and covering large distances in search of suitable host plants. Eggs of hawk moths are spherical, camouflaged green (cryptic), and are glued to the surface of the leaf, usually on the underside. In another, mostly day-flying, group of clearwing moths (Sesiidae), eggs are also laid singly, but are hidden in the crevices of the host plant or dropped on the ground near it. The young larvae must then burrow into the trunk or root where they develop. Some females, such as yucca moths of the Prodoxidae family, inject their eggs into the host plant tissue using an extended and sharpened tubular organ known as an ovipositor. Very frequently, however, eggs are laid in batches that are much more visible to predators and parasitoids (parasitic wasps and flies whose larvae develop inside immature moths). This may seem irrational, but saves females the energy deployed in flight, and minimizes the dangers they face from birds by day, and from bats by night. Caterpillars that result from eggs laid in batches frequently benefit from feeding in a group, which can more easily

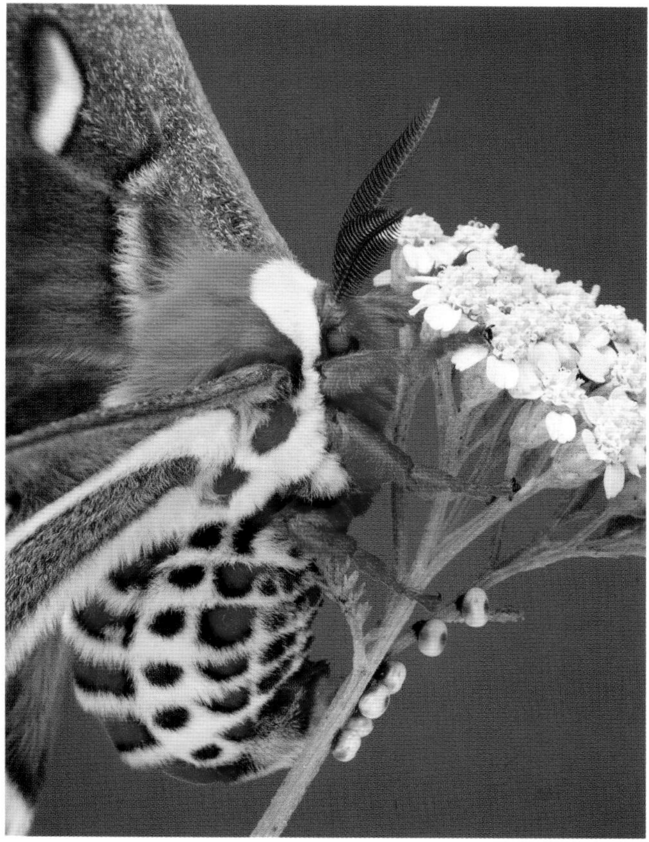

↖ The Fall Webworm (*Hyphantria cunea*) first lays its round eggs in batches of several hundred, then covers them with hairs from the tip of its abdomen.

↑ Eggs with a flat base, such as these of the Buff-tip (*Phalera bucephala*) adhere better to a leaf's surface. The black dots (micropyles) are where the sperm entered the eggs.

↗ The Cecropia Moth (*Hyalophora cecropia*) lays large hard oval eggs in small batches on a variety of host plants from over 20 families.

overcome plant defenses. The Ornate Bella Moth (*Utetheisa ornatrix*) and Io Moth (*Automeris io*), for instance, both lay a total of more than 300 eggs in batches of 10 to 50. As chemically defended caterpillars grow, they may either disperse and feed individually, or stay as a group and gain additional protection by sending a stronger warning signal to predators. In some cases, gregarious (group-feeding) caterpillars, such as the Eastern Tent Caterpillar (*Malacosoma americanum*) or Fall Webworm (*Hyphantria cunea*), also make a communal nest for additional protection, and, in these species, eggs are laid in especially large clusters; the female Fall Webworm moth lays 400 to 1,000 eggs in a single batch and dies after oviposition.

→ The Gypsy Moth (*Lymantria dispar*) lays batches of 300–500 eggs in a mass on a tree trunk and covers them with hairs, which protect them from predators.

↓ Ground Lackey Moths (*Malacosoma castrense*) lay overwintering egg masses in circles around branches and protect them with a clear secretion that hardens around them like epoxy.

COLOR, HAIRS, AND CHEMICAL PROTECTION

To blend with their surroundings, eggs may be translucent or cryptically colored to resemble a part of the host plant, or fungi, detritus, bird droppings, or any other feature of the landscape where they are laid. Minute as they are, moth eggs are a highly desirable source of protein for smaller predators and parasitoids, such as various ants, wasps, and minute pirate and big-eyed bugs.

Some female moths, such as Gypsy Moths (*Lymantria dispar*), cover their batches of eggs with hairs for protection, while Eastern Tent Caterpillar moths (*Malacosoma americanum*) use varnish-like secretions that harden and provide a shield for the eggs. While bird predation of Gypsy Moth eggs can be as high as 80 percent, evidence suggests that the eggs are unpalatable, as birds don't swallow them all at once, but eat them bit by bit, likely due to irritating hairs. The female Australian Processionary Caterpillar Moth (*Ochrogaster lunifer*) covers her egg masses with barbed tufts of hairs, while the South American silk moths in the genus *Hylesia* deposit urticant setae (irritant, hairlike bristles) over egg clusters that can cause allergic reactions in humans. Some tortricid moths, such as *Tortricodes alternella*, mask their eggs by using hardened structures on their abdomen to scrape dirt over the egg mass. Other species, such as the Ornate Bella Moth (*Utetheisa ornatrix*), lay brightly colored, poisonous eggs. A predator encountering them may taste one, but will leave the rest alone.

EGG SIZES AND INCUBATION TIMES

Partly as a result of different oviposition and survival strategies, but also depending on the adult size and species biology, the size of moth eggs can vary dramatically. This can be true even in very similar species of similar size: for instance, the Io Moth (*Automeris io*) and the Louisiana-eyed Silk Moth (*A. louisiana*), are so alike that they can form hybrids in captivity, but their egg size is significantly different. Among the largest moth eggs are those of the largest moths, such as the Asian Atlas Moth (*Attacus atlas*), whose eggs may measure ⅛ in (2.7 mm) in diameter and weigh 6 mg, while the egg of a tiny Erythrina Leaf Miner (*Leucoptera erythrinella,* Lyonetiidae) is microscopic.

While most eggs hatch within 2 to 14 days, depending on the species and the temperature, in some cases, such as the widespread genus of underwing moths (*Catocala*), eggs have a built-in delay mechanism, known as diapause, which ensures they are dormant through the winter. Eggs at a diapausing stage can be deceptive, however, as it is not the egg, but the fully formed caterpillar inside it that is most frequently overwintering.

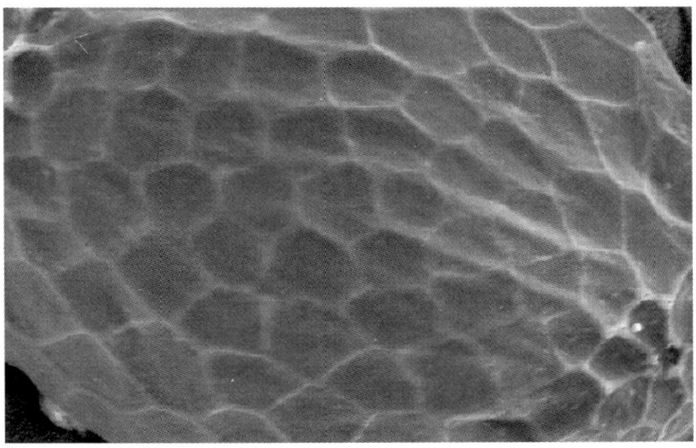

↑ The eggs of the Erythrina Leaf Miner (*Leucoptera erythrinella*) are among the tiniest—comparable in size to its host plant's leaf cells. While translucent when viewed under the light microscope (top), scanning electron microscopy reveals intricate details.

↓ Caterpillar of a Giant Owl Moth (*Brahmaea hearseyi*) hatching from an egg.

The ever-changing caterpillar

Moth larvae range in size from microscopic to 6 in (150 mm) or more in length and may be maggot-like, twiglike, leaflike, tentacled, hairy, spiky, or smooth—almost invariably more diverse than the stereotypical segmented grub with stubby legs of *Alice in Wonderland* or Eric Carle's *The Very Hungry Caterpillar*.

What they all have in common is their appetite and growth rate. Most pass through at least five stages (instars), during which they change in both size and appearance. The caterpillar of the Cecropia Moth (*Hyalophora cecropia*) weighs just 3 mg at birth but 5,000 times more when it reaches its peak growth within less than a month. Imagine a human baby growing from 8 lb (3.5 kg) to 20 US tons in one month on a strictly vegetarian diet. From a cryptic, brown, twiglike grub about ⅛ in (3 mm) in length, the Cecropia Moth caterpillar transforms into a green snakelike creature with yellow and blue protrusions along its body. The latter are not just decor: the spines help protect the larvae from vertebrate predators.

Io Moth (*Automeris io*) caterpillars undergo six instars as males and seven instars as females, and change greatly during their two-month-long development. Initially light brown (A), they soon acquire white stripes on a chocolate background, a coloration they maintain through the fourth instar (B), and become green or yellow in the fifth instar (C), when they disperse and feed solitarily until mature (D). Mature larvae display bright stripes, reminding predators of their venomous spines.

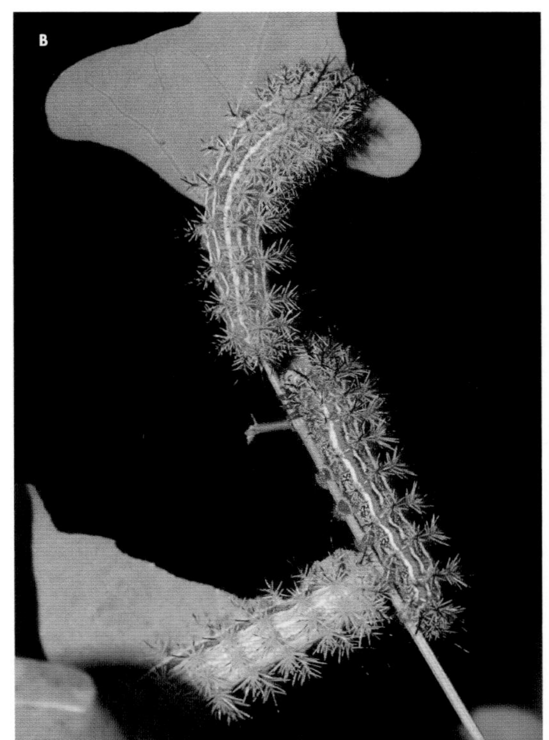
B

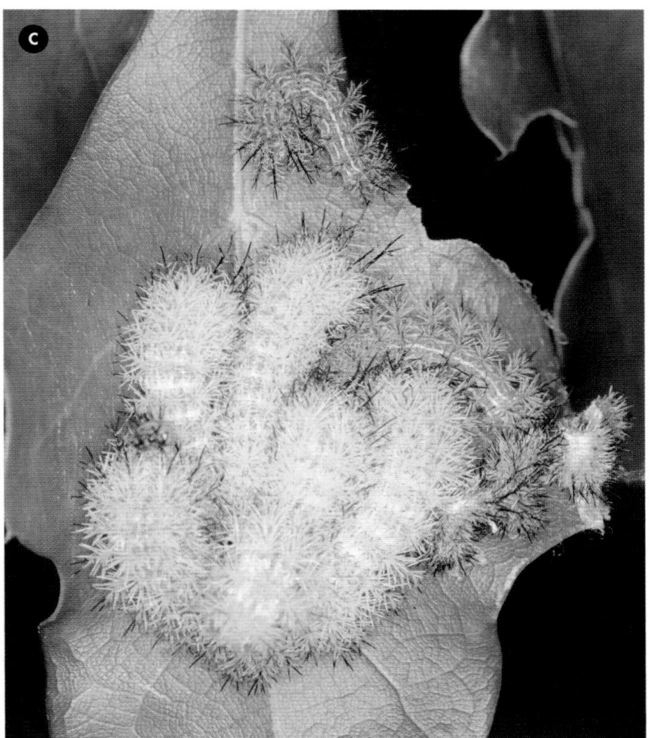
C

D

SENSING THEIR WORLD

Most caterpillars have six pairs of single-lens "eyes" (stemmata), but how well they see predators or other parts of their world is unclear. Rustic Sphinx (*Manduca rustica*) caterpillars, for example, will stop eating and "freeze" if they sense movement near them, but if someone approaching them stands still for a while, they will resume their feeding. Each stemma has a lens made of chitin (the same chemical from which the external skeleton of all insects is made), under which there is a crystalline cone. The stemmata also contain photoreceptors that transmit an external image to the optic lobe in the brain via seven axons—the nerve cell "wiring." The structure is much less complex than the adult moth compound eye, and it is commonly believed that caterpillars have minimal sight. However, research into jumping spiders suggests that simple eyes may work in synergy to produce a better image than one eye could achieve alone—similar to the way single-lens elements combine to make up the compound lens of a cell phone camera.

Their 12 eyes and 2 antennae, which communicate with the brain, enable caterpillars to sense and navigate their environment. They also have some form of

HEAD OF A CATERPILLAR

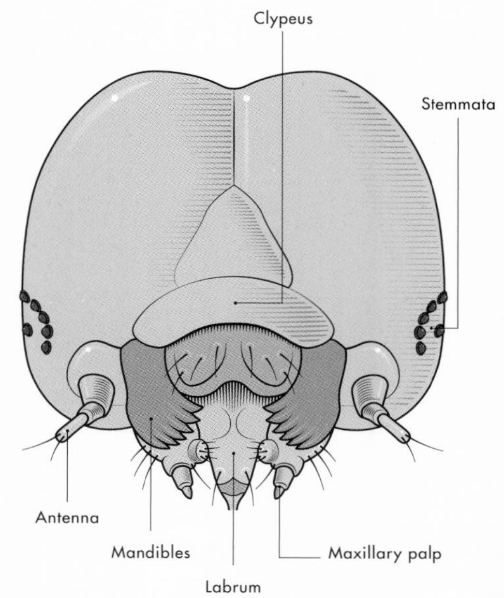

Protective capsule

Akin to a camera in a waterproof case, the caterpillar head is enclosed in a protective chitinous capsule. This changes several times as the caterpillar grows and molts, exponentially increasing in size.

FINELY TUNED TASTE

A simple experiment, such as placing an Ornate Bella Moth (*Utetheisa ornatrix*) caterpillar in one corner of a large cage and a rattlebox (*Crotalaria* spp.) plant in the opposite corner, soon reveals how effectively it will find a host plant. Caterpillars have two antennae responsible for their sense of smell, but use their mouthparts (maxillae and labrum) to taste and make a final host-plant choice.

For the Ornate Bella Moth, this choice is essential to its survival, as consuming the plant's bitter toxic compounds arms it with potent defenses against predators. Many caterpillars have evolved a finely tuned biochemical mechanism to use toxic plant compounds for their benefit.

← Top: A *Utetheisa ornatrix* caterpillar feeding on a rattlebox pod is protected by its host plant's toxic chemicals from a carpenter ant attracted by the plant's extrafloral nectar. Left: A *U. ornatrix* caterpillar feeds on developing seeds.

memory. Pavlov's dog-style experiments have proved that, by combining an odor with an electric shock, caterpillars of the Tobacco Hornworm (*Manduca sexta*) can be trained to avoid the odor and that memory persists through metamorphosis to the insect's adult stage. This suggests that whatever positive memories caterpillars acquire about their host plants may also pass to the adult moth, thus influencing the female's choices during oviposition. Like most insects, caterpillars are born with instincts that are quickly modified by experience. For instance, when a polyphagous caterpillar (one that can develop on many host plant species) begins feeding and becomes accustomed to one host plant, it becomes less inclined to accept another.

↑ The spines of a mature Royal Walnut Moth caterpillar (*Citheronia regalis*) are a formidable defense against birds.

CATERPILLAR LEG ARRANGEMENT

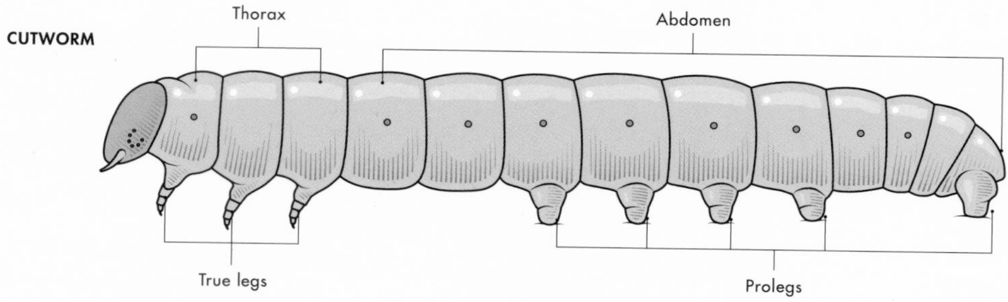

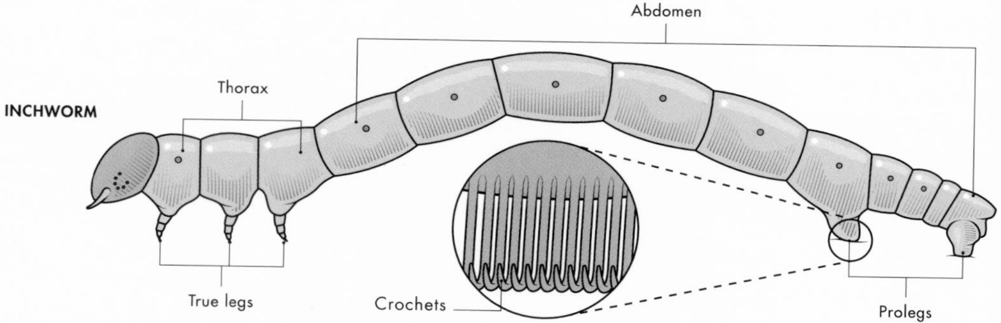

Abdominal segments

Prolegs on a caterpillar's abdominal segments assist locomotion. Depending on the family, they can vary in number and location. A cutworm (Noctuidae) has them on the 3rd through 6th and last abdominal segments, while most inchworms (Geometridae) have only two pairs—on segments 6 and 10.

MOVING AROUND

While caterpillars may appear multi-limbed, like all other insects, they have six true legs on the first three thoracic segments. In addition, however, they have a variable number of fleshy projections in the form of prolegs, which are equipped with crochets (sets of hooks) used for grabbing onto the host plant. Most caterpillars have five pairs of prolegs—four in the middle of the body and one at the end. However, the majority of inchworm (Geometridae) species have only two pairs on the last (tenth) and sixth abdominal segments; this helps distinguish the Horned Spanworm (*Nematocampa resistaria*) from the very similar-looking Curve-lined Owlet Moth (*Phyprosopus callitrichoides*), which has three pairs of prolegs, as both are found in the southern United States. Inchworms have an "inching" (earth-measuring) rather than a crawling gait, as do other caterpillars with a reduced number of prolegs, including the loopers, such as the Cabbage Looper (*Trichoplusia ni*).

In prominent moths (Notodontidae), such as the Puss Moth (*Cerura vinula*), found throughout Europe, the anal prolegs may be reduced and modified, giving them a very characteristic posture, with a raised tail end. Other caterpillars that tend to have reduced prolegs are those that move little but, instead, tunnel through plant tissue, hollow stems, roots, or inside a single leaf (as leaf miners do). Limacodid larvae, so-called "slug" caterpillars, have no crochets, but may have pads and suction cups instead, which enable them to stick to smooth upper leaf surfaces without having to spin a silk pad, although the wet secretion from silk glands helps them adhere to the leaf.

Most caterpillars make silk, which they use in various ways: to drop from a host plant when escaping a predator and then hoist themselves back again, to build a shelter or a communal nest, to attach themselves to a plant during molting, and to make a cocoon for the pupa. Silk proteins flow from paired silk glands—long organs running on both sides of the gut—and are combined into silk inside the spinneret. Unlike spiders whose spinneret is at the tip of the abdomen, a caterpillar's spinneret is on its head, protruding downward from between its labial palps.

↘ An inchworm, one of nearly 250 species in the genus *Eois*, navigating through the terrain in search of a place to pupate.

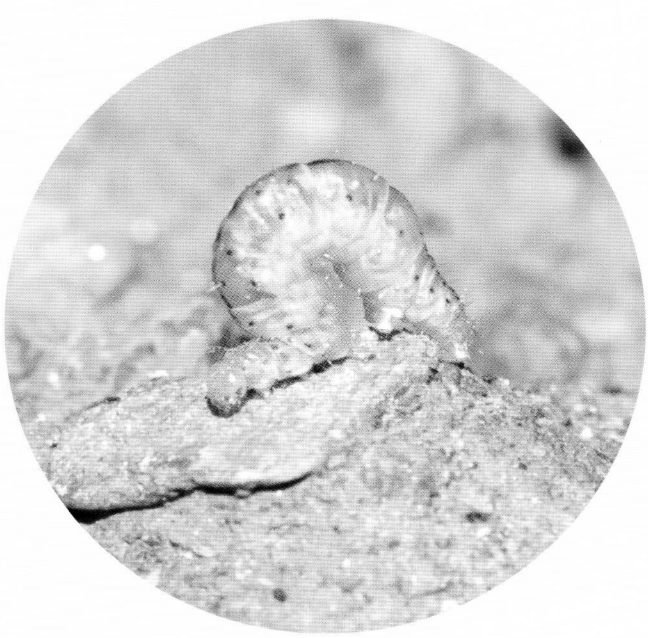

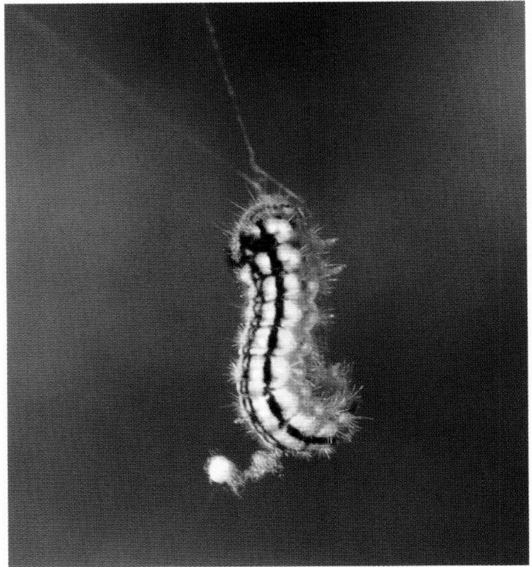

PARAGLIDING LARVAE

Some newly hatched (neonate) caterpillars crawl long distances, and Gypsy Moth (*Lymantria dispar*) neonates can also disperse by "ballooning." Aided by a silk thread they produce and their fluffy morphology, they paraglide through the air, assisted by the wind. As a result, siblings from the same egg mass may end up feeding far from each other as mature caterpillars, spreading the population over available food sources. If a group of caterpillars of the Laurelcherry Smoky Moth (*Neoprocris floridana*) consumes all the leaves on one tree, they will, like trapeze artists, swing on the wind suspended on long silk threads until they land on another host tree.

← Caterpillars of the Laurelcherry Smoky Moth (*Neoprocris floridana*) will swing on a silk thread from a defoliated tree to find a new host plant.

↖ The Curve-lined Owlet Moth caterpillar (*Phyprosopus callitrichoides*) mimics a hanging dry leaf not only with appearance but also by swinging side to side imitating leaf movement with the wind.

↗ The Echo Moth caterpillar (*Seirarctia echo*) has a warning pattern similar to that of a coral snake.

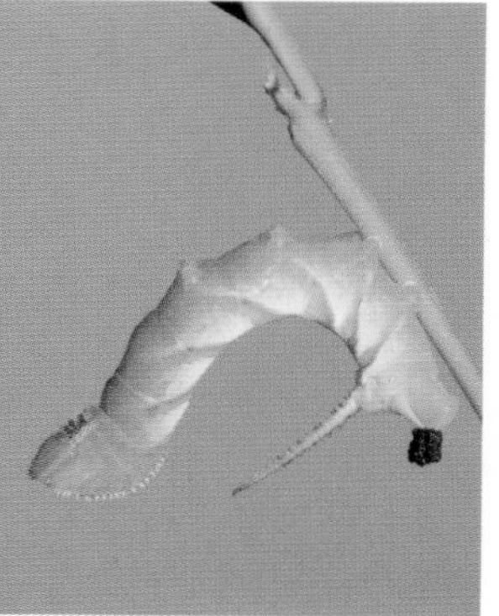
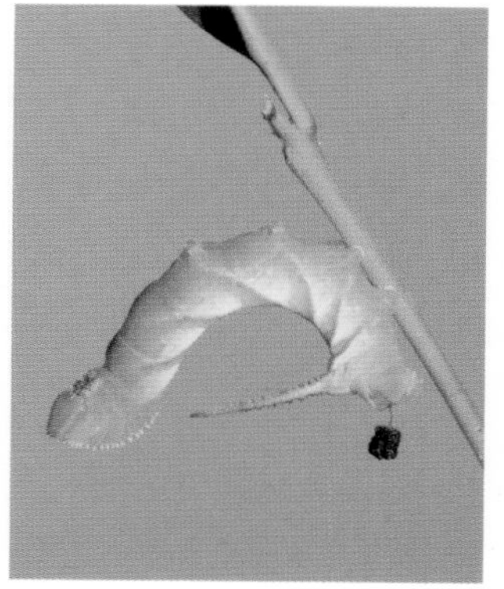
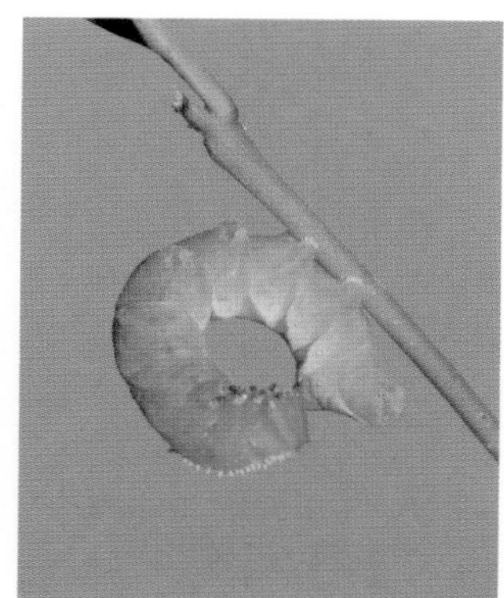
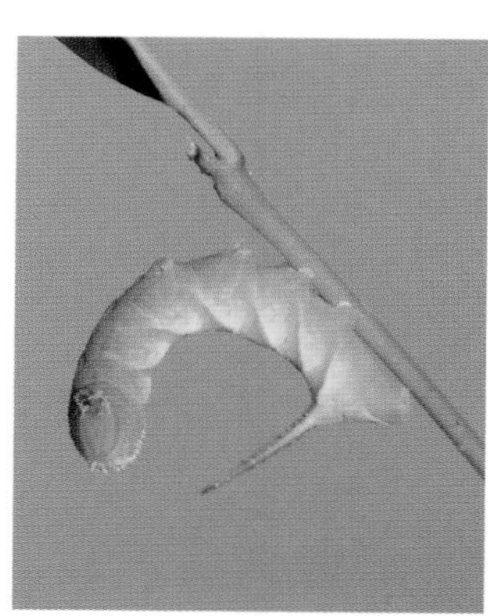

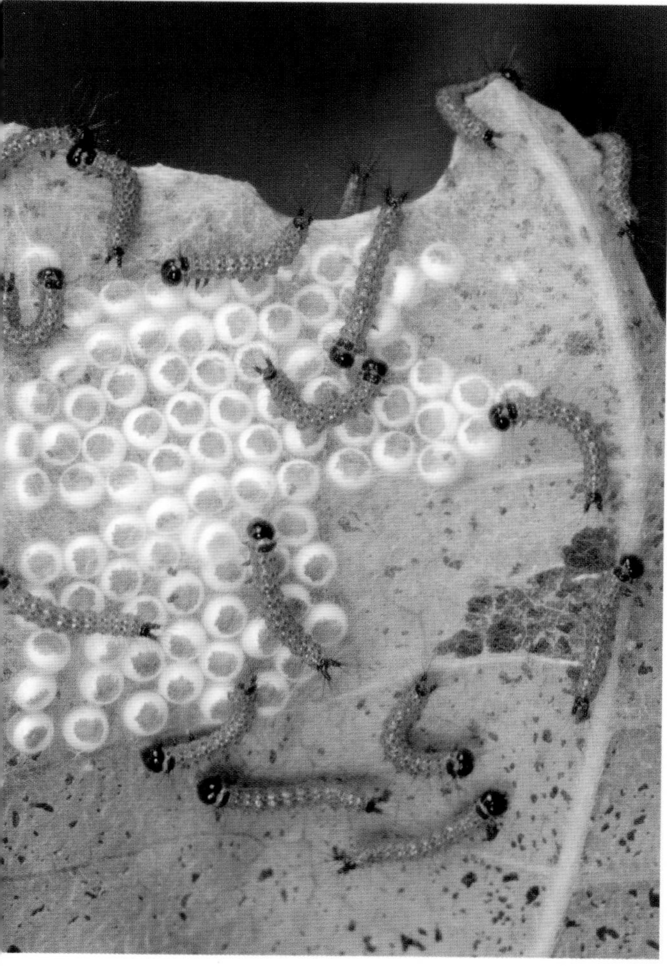

PROCESSING FOOD

When a new caterpillar hatches, chewing its way out of the egg, it frequently consumes the eggshell, then usually begins to feed on its host plant—first by scraping the leaf surface, then by eating through the whole leaf. Its flaplike labrum, a notched shield, will guide a leaf to the mandibles below. The labial and maxillary palps not only taste the food, but also direct it into the mouth. The insect's digestive system consists of the pharynx, esophagus, crop, midgut, and intestines, which open through the anus at the back where frass—the caterpillar's droppings—are excreted. Caterpillars frequently catapult frass away from

↑ A Rustic Sphinx caterpillar (*Manduca rustica*) defecates, in this case assisting itself with mandibles.

← Newly-born caterpillars of the Buff-tip (*Phalera bucephala*), like many larvae, eat some of their own eggshells after hatching.

→ A large Atlas Moth caterpillar (*Attacus atlas*) molts, shedding its skin (integument) as it grows.

themselves to avoid predatory wasps, which tend to locate the larvae by the smell of their droppings. Some caterpillars, however, such as the Asimina Webworm Moth (*Omphalocera munroei*), which lives on pawpaw (*Asimina* spp.) hosts, use their frass to make a shelter.

UPSIZING ADJUSTMENTS

As the caterpillar grows, its skin tightens and it has to change it, moving into its next instar. Normally, it will anchor itself with silk to the plant substrate and go into a passive state of molting that can last several days. Together with its skin, the caterpillar also upgrades its mandibles and trachea. The cutting edges of the mandibles are made of tougher, denser materials than the rest of caterpillar chitin, the fibrous substance that forms the outer layer of skin (cuticle) and of other external organs. Yet they, too, can wear out during feeding, which may explain why some species require more than six instars to reach their programmed size. Moth caterpillar diets vary greatly, from soft foods to tough, living hardwoods and silicon-rich palms and grasses that increase the wear and tear on mandibles. While the number of instars is species-specific, if fed on a substandard diet, a caterpillar may undergo additional molts. More frequently, however, a poor diet results either in death or smaller adult moths with poorer chances of survival. Well-fed caterpillars tend to live longer as adults, and the females are more fertile.

Sometimes the number of instars is directly related to the sex and size of the caterpillar, even within one species. For instance, caterpillars of the male Io Moth (*Automeris io*) molt six times, while females molt seven times. This allows female caterpillars to grow larger, resulting in adults that are twice as heavy as the males. Even though it takes them longer to develop, they emerge from pupae with more than 300 fully formed eggs in their abdomen. While being larger has certain benefits for males, too, they need to arrive first to locate females just as they are emerging from pupae to ensure they find a mate.

As in the Io Moth, in most Lepidoptera species, sibling males will develop as caterpillars and emerge a little earlier than their female counterparts. There are, however, some exceptions: female caterpillars of the Ornate Bella Moth (*Utetheisa ornatrix*), for example, always develop slightly faster than their male siblings, as male caterpillars in this species are collecting and storing defensive compounds for self-defense and pheromone production, which are also passed to the female during mating, adding to her own and progeny defenses. Unlike Io Moths, Ornate Bella Moths also feed as adults on nectar.

In all caterpillars, the rate of development depends on both the quality of the diet and on temperature: a drop from 73°F (23°C) to 59°F (15°C) can double the development time.

Pupae

Toward the end of its final instar, the caterpillar begins to prepare for a radically different pupal stage. Each species has its own strategy, but all have the same goal—to undergo the internal changes necessary for creating an adult moth.

PUPATION STRATEGIES

When fully mature, caterpillars shed excess fluids from their gut and change color. Species such as certain hawk moths (Sphingidae) and some genera of silk moths (Saturniidae) then crawl away from their host plant, burrow in the soil, and build an underground chamber; other species hide under rocks. Leaf rollers pupate within the rolled leaf where they developed as larvae, but leaf miners and stem borers may leave the protection of the plant tissues for the first time since entering them as microscopic neonates, and spin cocoons outside. Ultimately, when the adult emerges from its pupal case, and from any shelter used during pupation, it will need space to crawl out and suspend itself from a leaf, a branch, or a rock, and allow its wings to expand.

SPINNING SILK

Every moth caterpillar becomes a pupa, and many (but far from all) species, spin a cocoon—a silk sac that protects pupae during this vulnerable stage. While the pupae of closely related species may be similar in appearance, cocoons are generally highly diverse. Some species hang their cocoons from tree branches, while others hide them away by pulling a few leaves together. The pupa attaches itself to the inside surface of the cocoon by the cremaster—hooks on the last segment of the pupal abdomen. The cremaster is useful when the adult ecloses (emerges) as it enables the moth to break out of its chitin case without dragging it behind. In butterfly pupae (also known as chrysalides or chrysalises), the cremaster is often used to hook onto and suspend from a silk pad that a caterpillar makes before pupating. Some moth caterpillars also lay down a silk pad instead of spinning a full protective cocoon.

SERICULTURE—STOLEN SILK

For millennia, humans have cultivated the bombycoid Mulberry Silkworm (*Bombyx mori*) as well as saturniids such as the Temperate Tussar Moth (*Antheraea pernyi*), to produce a luxurious fabric from the silken threads the caterpillars spin into protective cocoons before pupation. While the first records of domestication come from China, wild silks from a variety of species have long been produced in many other countries.

While *B. mori* does not exist in the wild, it is so closely related to *B. mandarina*, a naturally occurring, mulberry-feeding silkworm, that the two readily mate in the laboratory, producing fertile lineages. *Antheraea pernyi*, used for tussar silk production, can live in the wild and may be the result of artificial selection from *A. roylei*, found in the Himalayas and Southeast Asia, as the two species mate and produce healthy offspring, despite different chromosome numbers.

In sericulture, except for rearing stock, moths are not allowed to emerge, as this damages the cocoons, which are boiled to soften their fibroin protein, making it easier to unwind the silk into strands. Each cocoon of *B. mori* is composed of a single strand of silk up to 3,000 ft (more than 900 m) long, but, because silk is so light, it takes up to 3,000 cocoons to produce just 1 lb (0.45 kg) of raw silk.

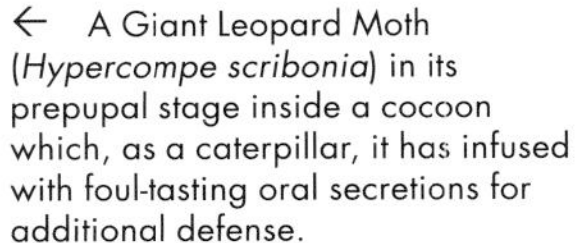

← A Giant Leopard Moth (*Hypercompe scribonia*) in its prepupal stage inside a cocoon which, as a caterpillar, it has infused with foul-tasting oral secretions for additional defense.

→ The Salt Marsh Moth caterpillar (*Estigmene acrea*) (top) constructs its cocoon from its own hairs, woven together with its silk.

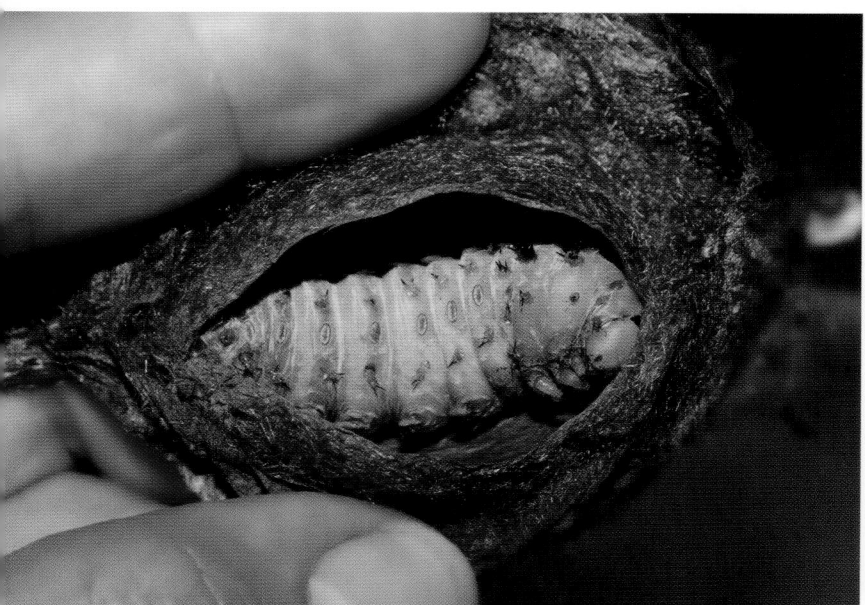

COCOON PROTECTION

Pupa is the moth's most vulnerable life stage as it has few defenses, unless it is toxic as a result of chemicals sequestered by the caterpillar. A cocoon offers crucial protection, but the degree of protection varies. *Antheraea* moths, such as the Japanese Silk Moth (*A. yamamai*) or in North America, the Polyphemus Moth (*A. polyphemus*), spin a very strong silk cocoon that can survive considerable mechanical damage and is an effective defense against squirrels and birds; only a very sharp blade could cut through its thin, but tough walls. These silver-colored cocoons are firmly attached with silk threads to bare branches of oak trees to ensure they stay in place even after a tree drops its leaves.

The Cecropia Moth (*Hyalophora cecropia*) goes a step further and spins a double cocoon: a loose outer one and a tightly woven inner one. The same is true of some crambid moths, such as the Erythrina Borer (*Terastia meticulosalis*) and the Erythrina Leaf Roller Moth (*Agathodes monstralis*). Silk cocoons, such as those of many tussock moths and tiger moths (Erebidae) and also flannel moths (Megalopygidae), are frequently enhanced with caterpillars' hairs. The hairs can be highly irritating to predators that attempt to break into the cocoons and gorge on protein-rich pupae. Additional secretions, such as caterpillar regurgitations, saliva, and frass, and foreign building materials such as sticks, rocks, or leaf particles are frequently employed to strengthen, protect, and hide the cocoons. Species such as the Diamondback Moth (*Plutella xylostella*) in the family Plutellidae create cocoons made of silk netting with large holes. As similar cocoons are found in some other tropical families, the strategy may have evolved to keep pupae dry in very wet conditions; by contrast, tight cocoons can prevent desiccation of pupae in dry conditions.

↖ A Cecropia Moth (*Hyalophora cecropia*) at its prepupal stage, which lasts more than ten days, shelters inside a double-layered cocoon (here cut open). The looser outer layer attaches to substrate; together with the tighter inner layer, it protects it against predators and parasitoids.

↑ The cocoon of the Gray Furcula Moth (*Furcula cinerea*) is camouflaged to resemble a swollen stem or branch of its host plant.

→ Leaf roller moths (Tortricidae) frequently develop as larvae and pupate concealed inside a rolled leaf held together by silk threads.

HIDING OR WARNING PREDATORS

Rather than spinning cocoons, many moth species protect themselves at this vulnerable stage by making their pupal shell of stronger, thicker chitin, similar to the exoskeletons of many beetles. These species include those of the genus *Eacles*, and many hawk moths, such as *Manduca* spp., which bury themselves in underground chambers, reinforcing them with saliva. Others, such as *Spodoptera* moth caterpillars, the infamous armyworms that damage vegetable crops, pupate in leaf litter beneath the host plants they feed on. Such noctuid pupae also can be found under rocks or hidden in the soil.

Toxic colorful caterpillars can sometimes turn into equally aposematically colored pupae, whose bright colors notify predators: "Don't eat me, I am poisonous." Such pupae may be protected by a translucent cocoon made of sticky silk and by the toxic host-plant chemicals sequestered (collected and concentrated) by the caterpillar. In addition to these forms of protection, Polka-dot Wasp Moth caterpillars (*Syntomeida epilais*) also add irritant caterpillar hairs to their cocoons and pupate in groups. In these cases, caterpillars that sequester toxins from their host plants may pass their chemical defenses not only to pupae, but also to the adult moths and even eggs.

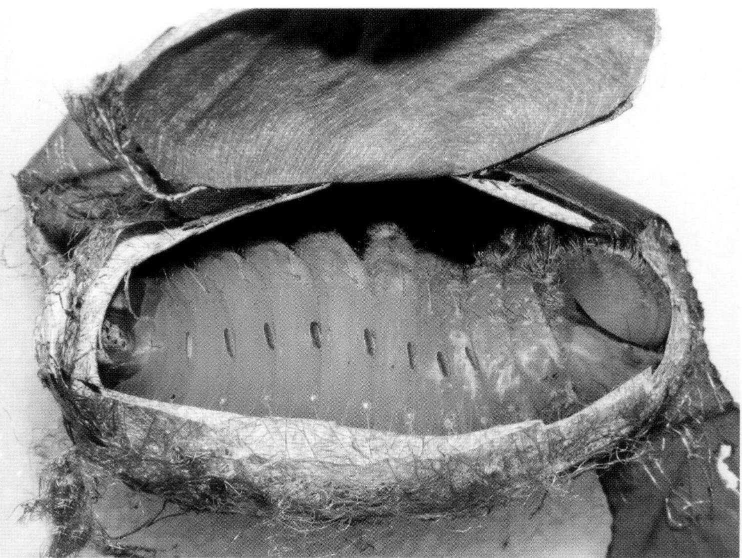

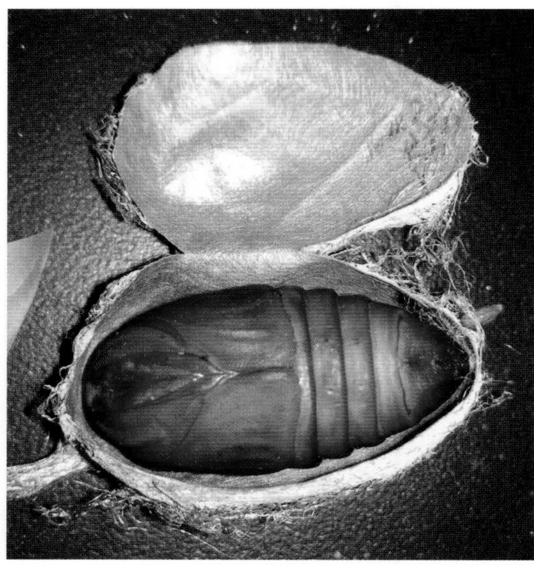

The developing pupa

When a cocoon has been spun, but pupation has not yet occurred, the caterpillar within it may have changed color and be smaller as a result of shedding food from its gut, but otherwise will react normally to light and other irritants. It will also still retain its ability to crawl and spin silk. If, at this stage, an incision is made in a cocoon, the caterpillar will repair it from within, and if a cocoon is removed, it will attempt to make another

↖ ↗ Prepupa (left) and pupa of the Polyphemus Moth (*Antheraea polyphemus*) inside a tough, silky cocoon, which shelters it from the elements and predators during harsh winter months.

↓ Mature sphinx moth caterpillars, such as this Tomato Hornworm (*Manduca quinquemaculata*), burrow into the ground or hide under rocks, where they pupate without making a cocoon.

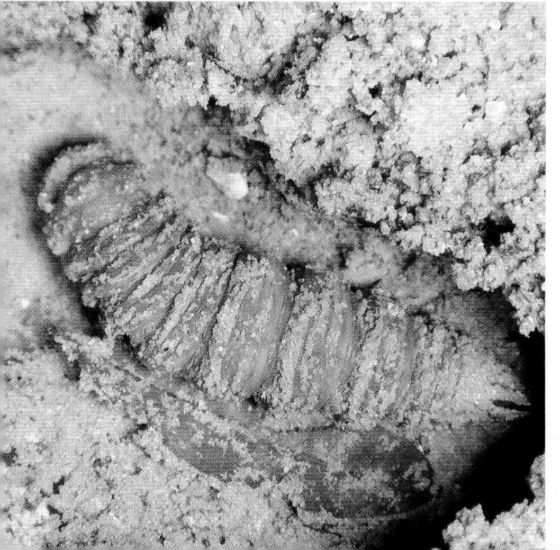

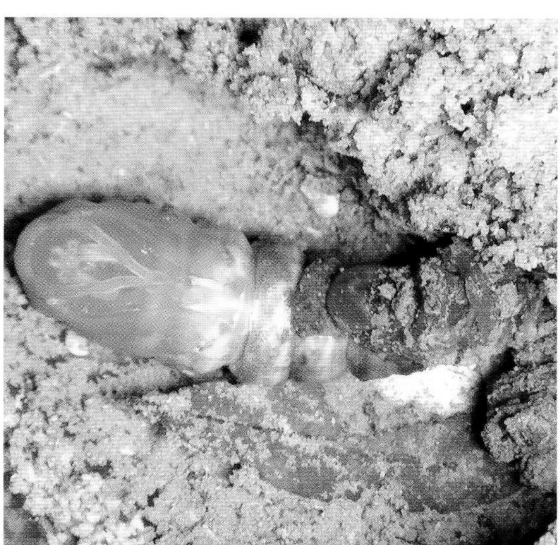

one, though it may be much less durable and complete, as its silk has been spent and time to pupation is ticking.

Next, the caterpillar turns into a prepupa. This is the stage during which the wing pattern begins to develop. Metamorphosis has already begun, as the final instar caterpillar starts to grow adult organs, such as wings and male gonads (testes). At the prepupal stage, however, the caterpillar becomes inactive and enters the passive stage of metamorphosis; its visible reactions to outside irritants diminish, and it continues to shrink in size and change color.

The duration of this stage depends on the species and the temperature, as do all aspects of insect development. For the Io Moth (*Automeris io*), prepupal duration at room temperature is four to six days, while for the Cecropia Moth it is 10 to 13 days, but raising the temperature from 68°F to 86°F (20°C to 30°C) can reduce this time by half. In some moths, prepupa is the diapausing stage, when normal development is placed on hold, usually to survive harsh conditions. For instance, in the fall, when the Erythrina Leaf Roller Moth (*Agathodes monstralis*) or Southern Flannel Moth (*Megalopyge opercularis*) enter their overwintering generations in north Florida, they remain in the prepupal stage inside the cocoon for several months, and pupate only two to three weeks before emergence.

Once the prepupa has transformed its caterpillar organs into the precursors of organs of the future adult moth, it becomes a pharate pupa, but retains its caterpillar-like outer appearance. Pharate pupae can be recognized by their paler and thinner caterpillar skin, which has now separated from the pupal cuticle underneath and is about to be shed. Like a snake slithers out of its old redundant skin after the new one is produced underneath, within three to ten minutes, the caterpillar skin splits and the new pupa—deformed, pale, and pliable—will wiggle out. Almost instantly, the new pupa starts to acquire its final shape and color during a sclerotization process called "tanning," in which the chitin fibers cross-link with other biopolymers (large molecules within the cuticle) and, as a result, the pupal case hardens and darkens.

Once the adult moth is fully formed, it breaks out of the pupal case (and, if inside a cocoon, out of the cocoon, too), spreads its wings, sheds its meconium (reddish fluid that contains by-products of the pupal metabolism), and flies away. In rare cases, moths never leave their cocoons. The wingless female of the Evergreen Bagworm (*Thyridopteryx ephemeraeformis*), for example, remains within her cocoon and the winged males mate with her through a special opening. The female then discharges her eggs and dies within the cocoon. As her larvae are highly polyphagous (feed on many plants), they usually have no trouble finding food.

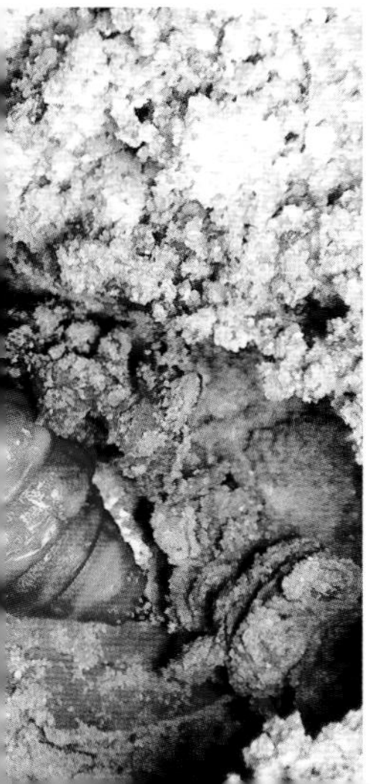

Metamorphosis

The caterpillar's transformation from a wormlike larva into a delicate-winged adult has fascinated humans for millennia. This seemingly miraculous change was considered a symbol of rebirth in many cultures; in Ancient Greece, psyche meant both "butterfly" and "soul." In fact, most insect orders are, like Lepidoptera, termed "holometabolous"—that is, they undergo a complete metamorphosis. Their larvae hatch from eggs and look quite different from the adult they become via the pupal stage. A smaller proportion of insects, including dragonflies, bedbugs, and locusts, are termed "hemimetabolous," as their wingless young (nymphs) resemble the adult form and skip the pupal stage.

A RESTRUCTURING PROCESS

Complete metamorphosis is usually viewed as a major change in form and function within one life span. Eggs, larva, pupa, and adult are the same organism with the same set of genes, but operate differently, under different disguises, and with different objectives. While the first two stages mostly focus on growth, pupae must survive adverse conditions and transform the organs of the caterpillar into those of the adult moth, whose principal duties are to mate and reproduce, dispersing the eggs.

As a caterpillar undergoes pupation, most of the organs allowing it to function are first reduced, and then restructured and renewed to produce the adult moth. During its life, a caterpillar also develops some organs, such as wings and testes, that become functional only in the adult. If the skin of a male caterpillar is translucent, the testes are visible—located dorsally toward the rear end and represented by a pair of (sometimes colorful) gelatinous organs. Spermatogenesis—the development of spermatozoa—begins to occur at the caterpillar stage.

Transformation

The Salt Marsh Moth (*Estigmene acrea*) lays 400–1,000 eggs, undergoes 5–7 larval instars (as the Salt Marsh caterpillar), and spins a cocoon out of its own hair and silk, where it transforms into an adult—completing all stages within 40 days.

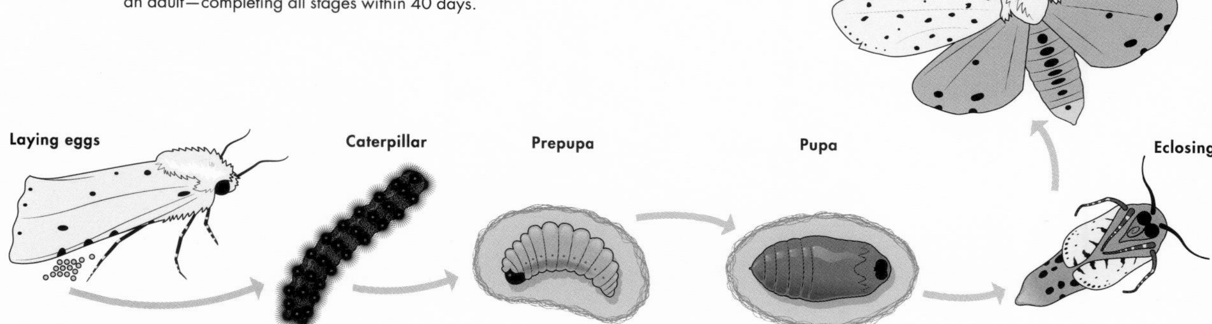

THE ROLE OF HORMONES

Metamorphosis is regulated in moths in a manner similar to the way in which development is controlled in humans—via hormones. The brain sends a signal to the prothoracic gland in the thorax to release ecdysone (the hormone responsible for molting), and instructs the corpora allata (another set of glands) to produce juvenile hormone. Together these two hormones regulate the production of the new cuticle in caterpillars and determine whether it molts into a larger caterpillar or becomes a pupa. Moths have played a crucial role in the understanding of metamorphosis as a phenomenon that all insects undergo. In the mid-twentieth century, 25 mg of ecdysone was first isolated from half a ton of pupae of *Bombyx mori* (the silkworm). Cecropia Moth (*Hyalophora cecropia*) pupae were used to determine the function of different organs mentioned above in hormone production and metamorphosis by surgically removing or implanting these organs.

The ovaries in females and genitalia of both sexes, however, develop during the pupal stage, as do many other important organs essential for the adult, such as the proboscis (straw-shaped mouthparts) and compound eyes.

Adult moth wings start to take shape within a mature caterpillar, tucked away under its skin between the second and third thoracic segments, but lack the beauty of those of an adult moth. They have no colorful scales and are very small—only about half an inch (12.5 mm) long in the large North American Luna Moth (*Actias luna*), although the developing wing veins can be clearly seen. At this point, the wings are similar to those of any other insect—clear membrane supported by thin veins—but they hold the single-cell precursors of future scales. Each of these cells will produce a single scale that takes shape toward the middle of pupal development and becomes pigmented even later. Throughout this stage, wings remain small, but expand when hemolymph (in insects, the equivalent of blood) is pumped through the veins as the adult emerges from the pupa.

→ Freshly emerged, first from its pupa, and then from its cocoon, a male Five-spot Burnet (*Zygaena trifolii*) hangs upside down to expand its wings before flying in search of a female.

Adult moths

A single-cell moth embryo inside a freshly laid egg becomes, within two months on average, a fully formed adult, waiting inside its pupal case for an opportune moment to eclose. When it first emerges, its wings are small and soft, so it needs a safe place to expand and allow them to harden before it can fly.

A SHORT LIFESPAN

Some moths emerge only in late afternoon to ensure they are ready to fly when it gets dark, while others appear to prefer different times of day. It is not clear precisely what triggers eclosion when pupal development is complete; it seems to depend on the individual species and its optimal conditions, such as springtime, rain, and warmer or cooler weather. Once out in the world, whether they are giants like the Atlas Moth (*Attacus atlas*), with a wingspan of some 10½ in (270 mm) or a tiny leaf miner, all adults are adept at finding a mate and reproducing. This needs to happen quickly: the life of most adult moths does not exceed two weeks, and many live only a few days.

WINGED WONDERS

Before taking its first flight, a moth needs to hang down, in order that gravity can help its soft, pliable wings to expand under the pressure of lymphatic fluid pumped from the thorax through its veins and spreading into wing membranes via diffusion. If its cocoon is attached to a branch, it will suspend itself directly from the cocoon; otherwise, it will need to crawl up from the ground and hang from a branch as its wings inflate. The moth's wing pattern is fully formed, but the scales are initially packed tightly together, so almost all of the wing pattern elements seem proportionately smaller.

The scales, which gave this insect order its name Lepidoptera (which is derived from the Greek for "scaly winged"), are marvels of engineering and nanotechnology. Thousands of tiny overlapping scales cover each wing, and each individual scale is a lightweight, almost hollow, ridged, chitinous plate. There are normally several different types of scales on every wing, varying in shape from paddle-like to hairlike.

↑ A large Atlas Moth male (*Attacus atlas*), a Southeast Asian species, takes flight.

→ The intricately shaped wings of the Twenty-plume Moth (*Alucita hexadactyla*) of the family Alucitidae span only $\frac{9}{16}$ in (15 mm).

← The striking Scarlet-bodied Wasp Moth (*Cosmosoma myrodora*), a day-flying wasp mimic, is found in the southeastern United States and Central America.

BRILLIANT COLOR

The shades of color we see on moth and butterfly wings are partly due to chemical pigments that absorb certain wavelengths of light and reflect others, while their brilliant iridescence is produced when light waves bounce off the surface of the insects' ridged scales and interact with reflected light. If, in the course of evolution, wings have lost scales and become transparent, as has happened, for example, in clearwing moths (Sesiidae), some species may resemble wasps or bees.

SAILING THROUGH THE AIR

A moth's wings have been compared to a boat's sails, with a membrane "fabric" supported by "masts and yards" of veins and controlled by the "crew"—the flight muscles in the thorax—which direct the airflow generated by the wing beat, while the nervous system captains the entire action. Before DNA-based classification, one of the few reliable ways to classify a moth was by looking at its wing venation. While scale covering and wing shape are features that change from species to species, the underlying "skeleton" of the wing is much more reliable and changes very slowly in the course of evolution.

The wing base and its attachment to the thorax is extremely complex. There are at least as many sclerites (chitinous plates) of different shapes and sizes as there are bones in our wrist, and their function is similar—multiple attachments to muscles are necessary to allow the wing to go up and down, move forward and back, tilt during flight, and fold against the body at rest. Moth wings are

↖ A highly magnified image of the iridescent wing scales of the Madagascan Sunset Moth (*Chrysiridia rhipheus*).

↑ A Tau Emperor (*Aglia tau*)—one of the few European saturniid species. Males fly during the day in search of females.

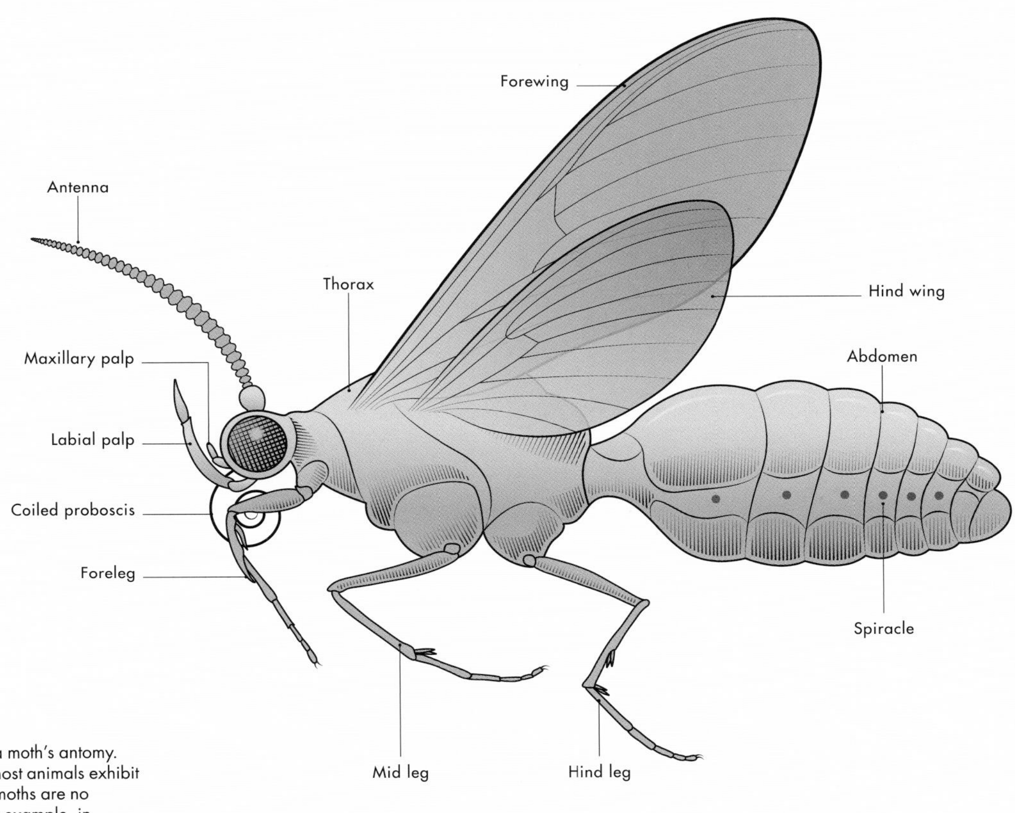

Segmented anatomy

Generalized diagram of a moth's antomy. From worms to humans, most animals exhibit body segmentation, and moths are no exception. Sometimes (for example, in the thorax, which consists of three fused segments) it is hard to see separate segments in a live animal.

much more intricate than any machine a human could engineer; they are versatile, living organs that the adult needs not only to travel but for many other functions, from thermoregulation to defenses and courtship.

THE INSECT BENEATH THE SCALES

Wings and scales are the moth's defining attributes, but otherwise Lepidoptera are quite similar to other insects. This is apparent when observing the wingless females in a number of moth families, which to the untrained eye are not easy to recognize as moths at all. The only unique moth feature is their mouthparts which, in most Lepidoptera, evolved into the proboscis, consisting of long, paired organs that zip together into a coiled straw shortly after the adult emerges from a pupa.

All the moth's organs are enclosed in an exoskeleton, just as in all other arthropods. Unlike humans, whose flesh and organs are relatively exposed, moths and other insects hide everything essential tucked away behind their chitinous armor, covered with scales and hairs. All the muscle attachments are on the inside too.

THORAX AND LEGS

The thorax is composed of the pro-, meso-, and metathorax. The moth's body is segmented like that of a caterpillar, but the segments are less noticeable on the thorax where the legs and the wings are attached. The legs may appear fragile, but they are adequate for a moth's needs—some species can even jump slightly while taking off from a flat surface. The claws are sharp and allow moths to hold on to the substrate effectively during periods of rest. The numerous sensilla (singular "sensillum"—a sensory organ containing nerve cells) throughout the legs suggest they play an important role both in sensing movement and in picking up other signals.

Occasionally, legs serve a different purpose beyond crawling and hanging. Some moths are spectacular mimics of wasp and bees, and their furry legs add to the illusion. In many hawk moths, sharp tibial spurs are an effective defense against vertebrate predators, while a number of moths use their hind legs to stridulate, producing sounds by rubbing them against special structures on the hind wings.

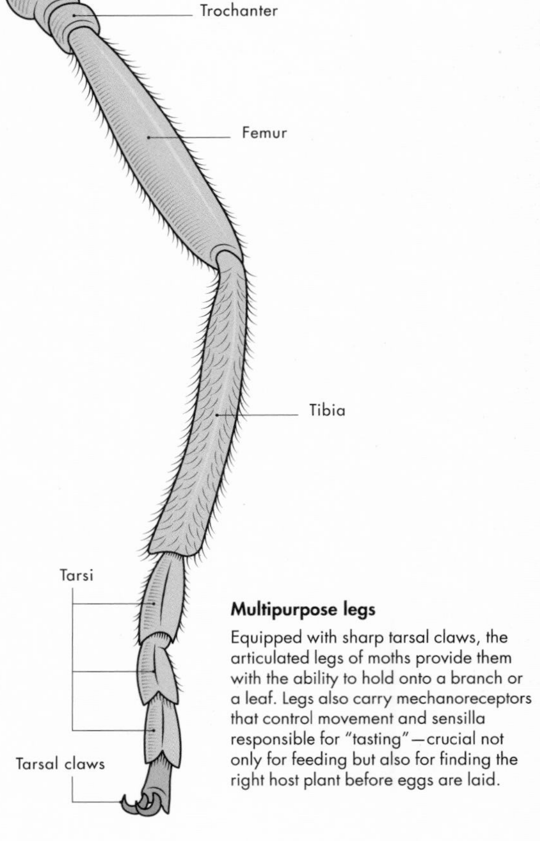

Multipurpose legs

Equipped with sharp tarsal claws, the articulated legs of moths provide them with the ability to hold onto a branch or a leaf. Legs also carry mechanoreceptors that control movement and sensilla responsible for "tasting"—crucial not only for feeding but also for finding the right host plant before eggs are laid.

ABDOMEN AND GENITALIA

The moth's abdomen is more obviously segmented than the first two sections of the body, the head and the thorax. Close to the point of attachment to the thorax, the abdomen may carry organs used for hearing and sound production. At the tip, the chitinous exoskeleton of several segments was modified in the course of evolution to form complex, articulated genitalic structures that are unique to each species.

As in caterpillars, each of the abdominal segments of the adult has spiracles—pairs of openings on the side through which air enters a branching system of tubes (trachea and tracheoles) that supply the inner organs with oxygen. Most of the organs present in caterpillars, though modified, are still there, although the adult has no silk glands. The male gonads develop at the caterpillar stage, but females have newly developed ovaries, often containing eggs and other reproductive organs.

↑ The furry abdomen of the Bedstraw Hawk Moth (*Hyles gallii*) is covered by modified scales. It encloses genitalia (that protrude during mating), and internal organs such as ovaries (or testes), Malpighian excretory tubules, and the fat body, which stores energy. Females may also carry in it fully formed eggs that are fertilized during ovipositing.

← The arctiine wasp moth *Horama panthalon*, found from Florida to South America, mimics a paper wasp (*Polistes* spp.), helped by the long colorful hairs covering its hind legs.

MOUTHPARTS AND EYES

The head, as in most animals, contains organs that allow the moth to sense the world as well as take in food, if the moth is a feeding species. It has antennae for smelling and orientation, multifaceted eyes for seeing by day or by night, and mouthparts. The earliest moths chewed on pollen, and some primitive moths have retained features that date from this evolutionary stage: instead of forming a proboscis, the same paired mouthparts are separate and short, and they have long maxillary palpi and mandibles. Most Lepidoptera, however, have a sucking pump connected to the proboscis that may also be equipped with sensilla for tasting and with hooks for piercing fruit, and many other adaptations, depending on diet. The proboscis may be reduced in non-feeding species or extended to three times the length of the body in certain hawk moths. Labial palpi, which moths use to sense food and to wipe their eyes, can also be very prominent.

On top of the head are three ocelli, simple eyes that are somewhat like those of the caterpillars, but unfocused, so they produce no image, although they likely help with orientation. Compound eyes, which occupy most of the head, consist of many, and in some species many thousands, of facets—the optical units known as ommatidia. Each ommatidium has a crystalline cone and is covered by the outside corneal lens. The double lenses in each ommatidium work like a simple telescope, enlarging whatever they perceive from far away, and, very often, the separate ommatidia all make their own image, which the brain then combines.

The brain of Lepidoptera is remarkably complex for its size, with dozens of different types of neurons and numerous compartments. In addition to processing the information from sensory organs, the brain sends signals to adrenal glands and, via hormones, governs many physiological processes, such as mating and egg production. In long-lived Lepidoptera, some parts of the brain can grow with experience, so, just as in humans, learning and not just instincts guide their actions. Additionally, nerve centers or ganglia are located in each of the body segments to process information and to produce faster local reactions.

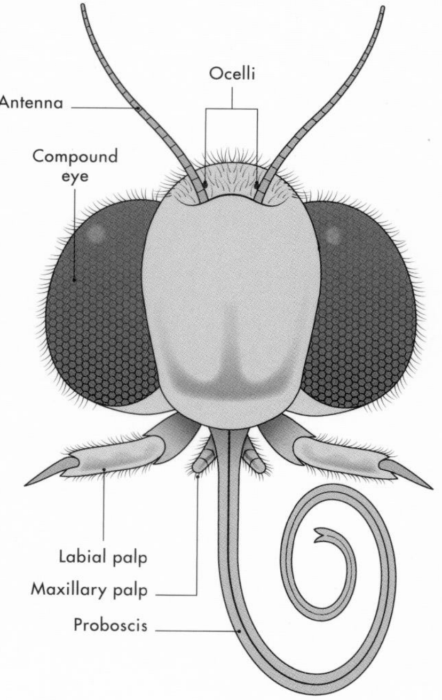

Eyes and mouthparts

The moth's compound eyes are typical of most insects allowing for 360-degree vision, while the mouthparts, uniquely for Lepidoptera, evolved into a straw-like proboscis adapted for drinking.

← Head of the Subflexus Straw Moth (*Chloridea subflexa*).

↓ Tiny white pollen particles adhering to the curled proboscis and labial palpi of the Giant Sphinx Moth (*Cocytius antaeus*).

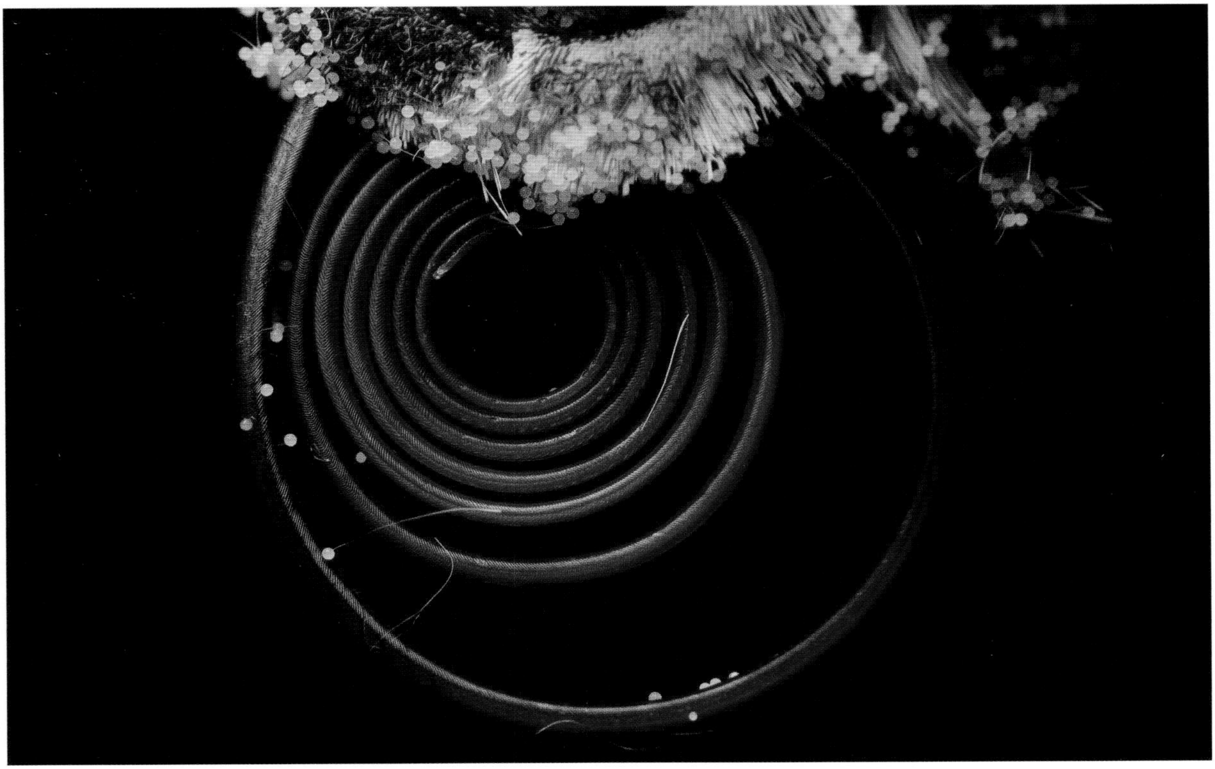

INTERACTIONS

Courtship and mating

A male moth's prime instinct is to court and mate. Males patrol their territory looking for females. While some diurnal (day-flying) moths use mostly visual cues, most moths are nocturnal, and their males are on high alert for the pheromones—airborne chemical signals—that females emit. At close range, many males also produce pheromones and "sing" ultrasound songs to woo a female. In some species, males lek—they gather in a group and perform aerial displays. Each species has its own strategy, and its own behavioral, morphological, and biochemical adaptations. Mating is a complex process that transfers not only sperm but also mixtures of other chemicals from male to female, who then uses the sperm and these nutrients gradually throughout her life.

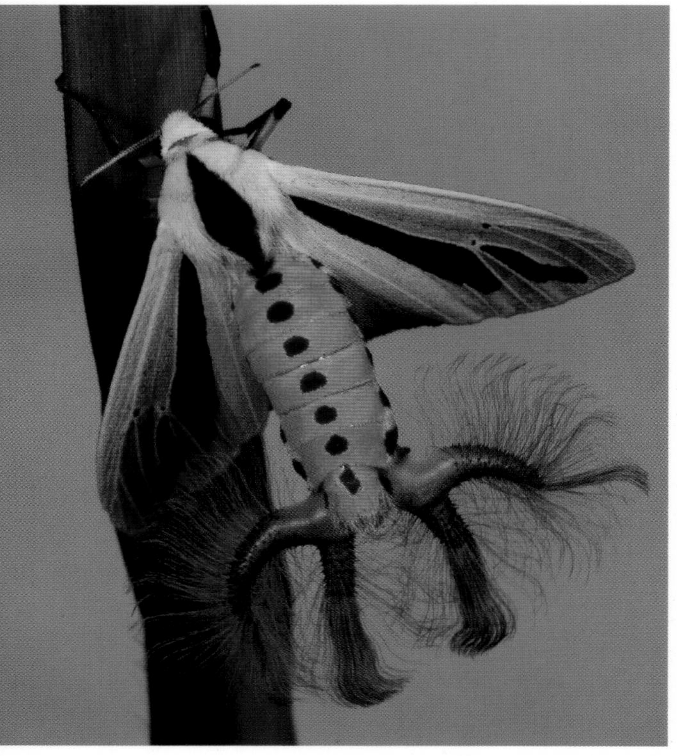

COMPELLING PERFUMES

Lepidoptera courtship begins with pheromones, produced in specialized glands and released via a variety of brushlike organs. Normally, the opposite sex detects them via antennae, which stimulate a behavioral response. Pheromones differ chemically according to species but must be volatile—consisting of molecules small enough to become airborne, creating a trail for a moth to follow. Once a mate is located, the moths may also emit "contact pheromones" that stimulate mating behavior during courtship. The pheromone mixture must be species-specific, at least locally, where the moth occurs, to avoid unwanted attention, which suggests that there are over 100,000 different pheromone mixtures plus many more of species not yet identified. Studies of some moth groups indicate that pheromone mixtures and host plant choices

← A colorful male Baphomet Moth (*Creatonotos gangis*) displays its large inflatable coremata, bristling with hairs. These organs unfurl from the abdomen to disperse pheromones to attract females.

have evolved together, although the relationship is far from linear. Among closely related species, the pheromone mixtures may differ by only a single compound or, more commonly, just by differing ratios of the same compounds.

In many species, such as the Codling Moth (*Cydia pomonella*), which is a pest on apples, pheromone-producing behavior intensifies when the female is close to its host plant. This may be true of many moths as it is advantageous to mate near the oviposition site and start laying eggs immediately. To release their seductive scent, some moths extrude organs called hair-pencils, which can be found in both males and females. For example, these are protruded and spread out from the tip of the abdomen in calling females of *Diaphania* crambid moths, while in *Tarsolepis* spp. (Notodontidae)—the tear-drinking moths from Southeast Asia—scented red hair-pencils are located close to where the abdomen connects to the thorax. Many tiger moths have additional inflatable sacs called coremata that are covered with scented hairs; the most spectacular variation of coremata is found in the Old World tropical genus *Creatonotus*, where these organs can be as long as the moth itself and are everted (turned inside out) when the moth is "calling" its mate. Male Ghost Moths (*Hepialus humuli*) release pheromones from special brushes on their legs, and some moths, such as the day-flying Palm Moth, *Paysandisia archon* (Castniidae), have scent (or androconial) patches on their wings instead, as do many male butterflies.

↑ About half of hawk moth species have pheromone-releasing brushes, such as this on a section of the abdomen of a male *Enyo* hawk moth in French Guiana. Normally, they are located in a pair of pockets that extend across the second and third abdominal segments. Scent is produced by cells at the base of the brush and is dispersed when the brushes are exposed. Similar brushes are found in other moth families, for example Notodontidae.

← A Walker's Moth (*Sosxetra grata*), an unusual noctuid in the subfamily Dyopsinae, found throughout the Neotropics, displays serrate antennae that enable it to pick up scent, and long hairs on the margin of its hind wing that may be involved in both defense and mating.

EXPERT DETECTORS

Many male moths, especially silk moths (Saturniidae) and some other bombycoids, have much more serrate (featherlike) antennae than the females whose pheromones they must detect. This gives their antennae a larger surface area that can hold more sensilla (sensory organs), enhancing their pheromone-detecting abilities. Males following the pheromone trail, sometimes from miles away, can react even to a single pheromone molecule. Once they locate a female, both sexes activate a cascade of responses. At first these are limited to search behavior but, when the male finds and physically touches the female, this produces a hormonal response that triggers the mating behavior. During courtship displays, which are sometimes prolonged, males may release short-range contact pheromones, literally showering females with them in a bid for acceptance.

Scent-detecting organs

Diagram of a single sensillum—a scent-detecting organ on a moth's antenna.

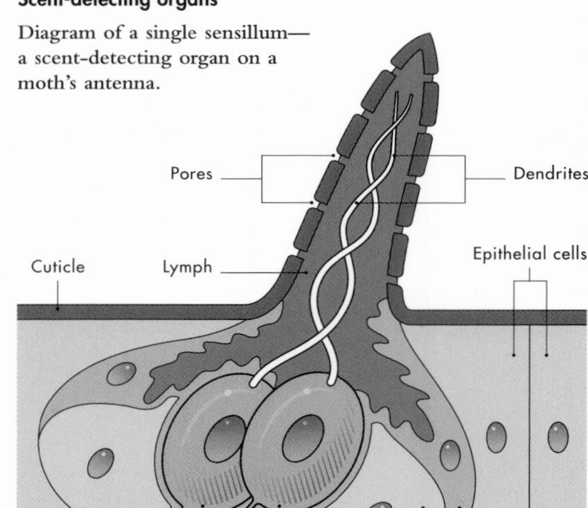

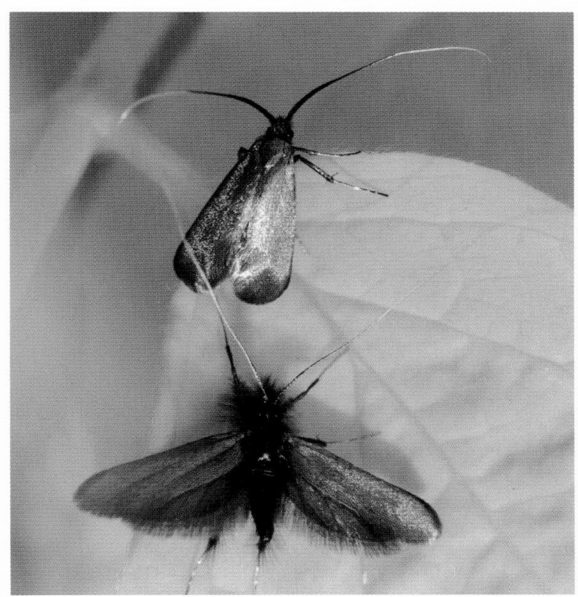

FATAL ATTRACTION

Certain spiders have evolved a hunting strategy that exploits the male moth's attraction to pheromones. The bolas spider (*Mastophora hutchinsoni*) produces pheromones during the night to attract males of either the Bristly Cutworm Moth (*Lacinipolia renigera*) or the Smoky Tetanolita Moth (*Tetanolita mynesalis*). Remarkably, the pheromone mixture changes throughout the night, depending on which species it is attracting. By day, the yellow garden spider (*Argiope aurantia*) attracts the diurnal males of oak worm and buck moths of the saturniid genera *Anisota* and *Hemileuca* by mimicking pheromones of these species. Like spiders, agricultural entomologists also replicate moth pheromones, but can only do so by using sophisticated chemistry techniques. They may use the resultant synthetic pheromones to monitor moths in fields and orchards, but to kill pest moths, the pheromones are usually combined with pesticides or with a sticky trapping glue. Pheromones sprayed in a field can also disrupt a pest's mating.

HOW RESEARCHERS STUDY PHEROMONES

Researchers take the following steps to identify the moth pheromones:

Step 1: To collect the moth pheromones, air is pumped through a glass container with unmated virgin females, and as it exits, the air passes through filters—narrow tubes filled with absorbent, chemically active material. Alternatively, chemicals can be extracted from the pheromone-producing glands of the moth.

Step 2: Gas chromatography and mass spectrometry are used to separate and identify chemicals by weight. They are then compared to a database of already identified chemicals.

Step 3: To discover which of the chemicals are pheromones, as they exit the gas chromatographer, they are directed onto a male moth's antenna attached to an electrode. The sensilla on the antenna react only to very few of the volatile compounds.

Step 4: To test for behavioral response, the moths are observed in an air tunnel—if they are attracted to the mixture, the pheromone chemicals have been correctly identified.

↖ Male Green Longhorn moths (*Adela reaumurella*) court females by vibrating their long antennae.

← The male Cecropia Moth (*Hyalophora cecropia*) has much broader feathery antennae than the female; their larger surface area allows for more sensilla—scent-detecting organs.

HOOKING UP, LITERALLY

Moth taxonomists spend much of their time examining the mating parts of the moths under a microscope while describing and identifying species. Together, male and female genitalia are thought to work like lock-and-key mechanisms. The way the genitals are configured is unique to each species, which prevents any interbreeding between species, as this would likely result in non-viable offspring. Once the "key" is in the "lock," a complex mixture containing not only sperm, but also additional chemicals useful to a female, is transferred in a form of a "package," called a spermatophore, during the mating process, which can last from minutes to hours.

In some cases, special spines called cornuti are transferred to the female and, while their function is unclear, it is thought that they may assist in future fertilization and prevent the female mating with other males. A female keeps the spermatophore inside her abdomen in an organ called the bursa copulatrix, from where spermatozoa swim through a narrow tube to a vagina to fertilize eggs before they are laid. Most moths have two openings, one for mating, the other for egg laying, so mating and fertilization are two separate processes occurring at different times. The number of spermatophores found in a bursa indicates how many times the female has mated.

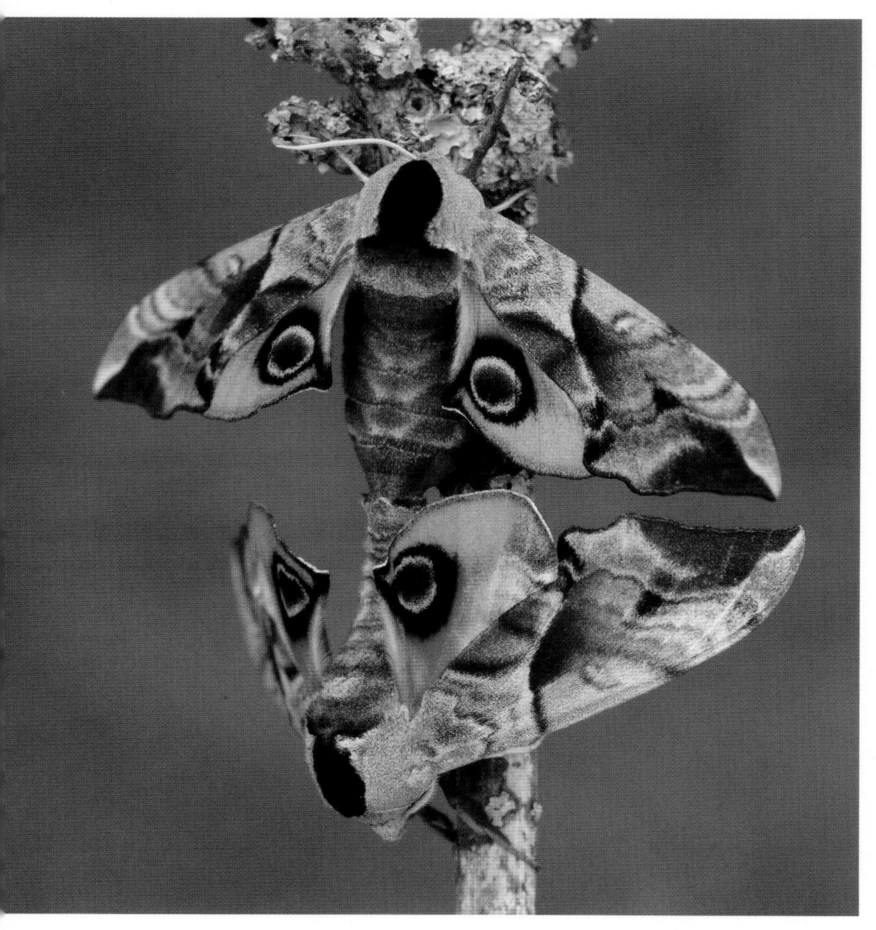

← A mating pair of Eyed Hawk Moths (*Smerinthus ocellatus*). The female (top) is usually the one holding up the pair with its strong legs during mating, while the male hangs passively downward.

↗ In a behavior called lekking, male Green Longhorn moths (*Adela reaumurella*) gather and perch on vegetation, then fly up together in a display designed to attract females.

OTHER COURTSHIP SIGNALING

In addition to pheromone signals, many moths produce sounds related to courtship. Frequently, complex species-specific ultrasound signals have evolved, and a male may serenade a female before mating occurs, while a female may respond with sounds of her own, signaling acceptance. In butterflies, a group that evolved from moths about 90–100 million years ago, wing patterns play a major role in species recognition for courtship and mating. The visual clues in butterflies have largely replaced the long-range pheromone signaling found in moths. Likewise, in some diurnal colorful species, such as the Palm Moth (*Paysandisia archon*) and other moths in the family (Castniidae), females have lost their pheromone glands, and males, who use visual cues to locate females, produce only short-range pheromones to woo females. Research into color dimorphism (the presence of two color forms) in the Wood Tiger Moth (*Parasemia plantaginis*) males in Europe, suggests that, while it may increase bird predation, it may also encourage females to select them. In several species, such as the diurnal Green Longhorn (*Adela reaumurella*), known for its long antennae and shiny wings, males gather in large groups in a mating area to perform display flights to attract females—this is known as lekking. This lekking behavior is commonly found in other insects but is uncommon in nocturnal moths. However, in crepuscular male ghost moths, lekking may also be a bid to enhance their silver-white coloration so that females can see them more easily at dusk. Both examples suggest that visual cues, and not just pheromones, may be much more essential in moth mating than is commonly thought.

Feeding as adults

Like most butterflies, many moths suck nectar from flowers, while some feed on other foods such as tree sap, honey, rotting fruits, or even tears. A large number of moth species, however, need no food as adults because, in their larval form, they've collected enough resources to enable them to mate and reproduce. Non-feeding species, such as all silk moths (Saturniidae), may not even have a functional proboscis—the extended mouthpart other species use to suck fluids. Once their resources are spent in mating and egg laying, they will die. Adults that are equipped to feed may live two to three weeks, and occasionally close to two months. Although experience may prompt some brain development, moths do not grow as adults. Their size is an interplay of genetics and the individual's conditions as a caterpillar, such as quality of food.

← The Nine-spotted Moth (*Amata phegea*), found mainly in southern Europe, sips nectar from a lavender flower head.

↗ An Ailanthus Webworm Moth (*Atteva aurea*) feeds on garlic chives blossoms.

→ The Sugar Cane Borer Moth (*Amauta cacica*) feeding on a *Heliconia* flower in Costa Rica. Its larvae bore into the roots of *Heliconia* spp. and also plantains.

SUGAR AND FLUIDS FOR LONGEVITY

All moths, whether feeding or not, have a "fat body," a fat-storing organ inside their abdomen that functions somewhat like a liver. The fat body is responsible for producing sugar and protein, and releasing them into hemolymph (in insects, the equivalent of blood) to fuel the moth's energy needs. In feeding moths, acquired nutrients will help them live longer and be more active. In studies, the Spruce Bud Moth (*Zeiraphera canadensis*) and a number of other species have lived on sugar water for about three weeks, and some, such as the Ornate Bella Moth (*Utetheisa ornatrix*), for more than six weeks. Metabolism greatly depends on temperature, and thus, with rising temperatures, moths use more energy just to stay alive; a moth placed in a refrigerator at a temperature of 41–50°F (5–10°C) can live longer than the same moth at room temperature. While studies suggest that access to sugar is very important in moth longevity and fertility, access to water is even more crucial. Nectar-feeding moths that can live on sugar water for up to two months, can live without sugar for a month; without water, however, their life span is no longer than a week.

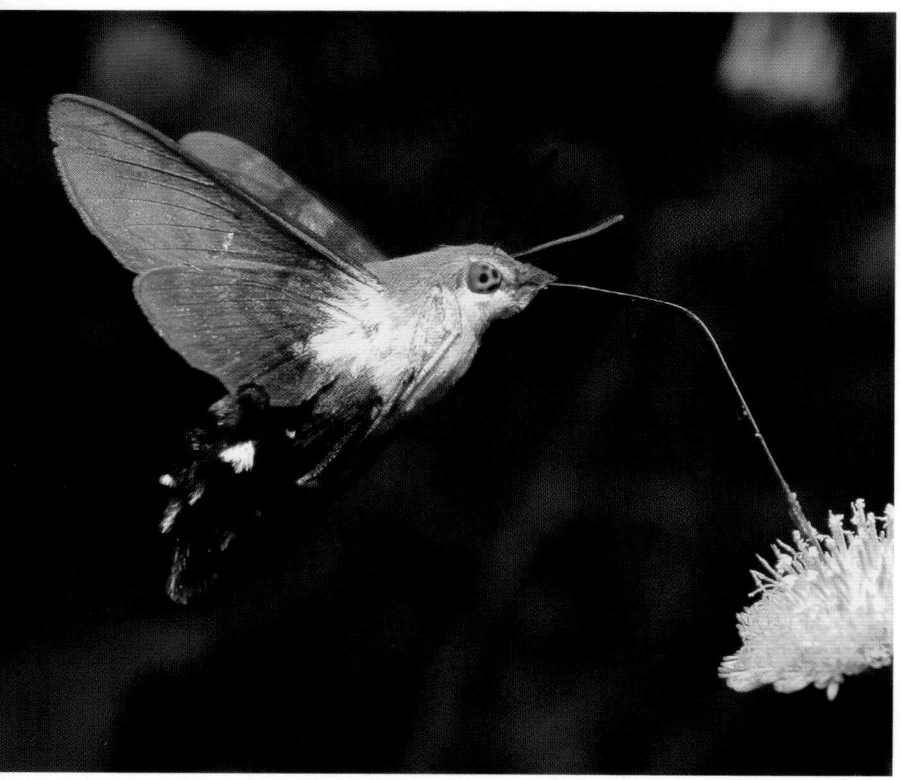

← The Humble Hummingbird Hawk Moth (*Macroglossum bombylans*) extends its long proboscis into a flower head to feed on nectar in Yunnan, China. The species gets its common name for its humming noise and hovering behavior.

→ A day-flying tiger moth in the genus *Gnophaela*, which includes five similar species in North America, is feeding here on tansy ragwort flowers in Oregon. Their distinctive black-and-white coloration deters predators.

FEMALE FERTILITY AND MALE ENERGY FOR MATING

Moths that live longer can produce more offspring. While a female may emerge from pupae with fully formed eggs, ready to be laid, in many feeding species, the eggs need to mature first, and more will mature the longer she can live. In the Spruce Bud Moth, for instance, the peak of egg maturation is on the tenth day of its life. Research suggests, however, that this can be achieved only if the female is given a 5 percent sugar solution. Larger females tend to live longer and can lay more eggs, but only if they have adequate nutrition.

Both sexes seek nectar, and feeding males need energy from sugars to find females and to restore their reproductive potential after mating, so that they can mate several times. This is partly because mating takes its toll. Male moths sacrifice their own longevity to invest instead in their future offspring by transferring nutrients to the female, as well as sperm during mating, although the effects vary according to species. In one study of the Copitarsia Worm Moth (*Copitarsia decolora*), males that mated lived for a shorter period than those that did not mate. The Western Avocado Leaf Roller Moth (*Amorbia cuneanum*), however, appears to be able to mate at least once without any noticeable loss in longevity. The effects of mating on moth fitness and longevity may be different because of the variable amounts and types of nutrients that males of different species transfer.

SUGARS AND HELPFUL POISONS

Nectar is the food of choice in many moth families, such as hawk moths (Sphingidae), sky-pointing moths (Crambidae), owlet moths (Noctuidae and Erebidae), butterfly moths (Geometridae), and more. However,

the nectar composition requirements can differ between species. For instance, species that hover over flowers, such as hawk moths, need more carbohydrates, while others may need more amino acids. Attraction to flowers may also correlate with the time of the night (or day) when the moths are active, and whether host plants for egg laying are nearby, which makes foraging more efficient.

In some cases, the moth's host plant is also its source of nectar, which is a fortunate, but less common arrangement. Some forms of nectar are toxic and, as with caterpillars, only a specialist moth can feed on it. The Police-car Moth (*Gnophaela vermiculata*), for instance, has no trouble feeding on the flowers of the tansy ragwort (*Senecio jacobaea*), which contain toxic compounds called pyrrolizidine alkaloids (PAs). Feeding on such toxic compounds can add to a moth's defense, and some moths look for them in nature, not only from nectar, but also from wilting tissues of toxic plants. For example, the wilting leaves of heliotropes and other plants that contain alkaloids, attract colorful, chemically defended tiger moths, which seek PAs to enhance their defense and reproduction potentials. In some species, only males are attracted to such plants but in other species only females seek them out. In Florida, the Edwards Wasp Moth (*Lymire edwardsii*) feeds on PA-containing tissues of wilting dogfennel (*Eupatorium capillifolium*). For these moths, PAs are a source of nitrogen, which is uniquely accessible only to species that can detoxify the alkaloids.

NOCTURNAL FEEDERS

Fields of Spanish needles (*Bidens alba*) attract insects both by day and by night. The yellow disk of these flowers is highly visible to diurnal insects, but the peripheral white flowers that frame the disk are more visible to nocturnal insects. Many small, white, fragrant night-blooming flowers target exclusively nocturnal insects as pollinators, and moths are among them. When searching for nectar, moths can detect their floral volatiles (complex multifunctional signals), and may have innate preferences but quickly learn to associate reward with specific smells, so their instincts are modified by experience.

Some rare flowers, such as orchids, which require specialized pollinators, may produce volatiles that target specific hawk moths. Certain orchids may even have unique chemicals in their nectar (in addition to sugars) that further reward these moths, and these plants grow long spurs, with nectar available only at the deepest point, thus limiting the number of visiting species to those with a long proboscis. This increases the chances

↑ An Elephant Hawk Moth (*Deilephila elpenor*) in flight, feeding from a honeysuckle flower at night in Oxfordshire, England.

BIOENGINEERING MARVEL

The moth proboscis is a marvel now studied by bioengineers. There are numerous applications for a smart straw that can coil and uncoil, and collect vast amounts of liquids without expending much energy. It consists of two separate paired mouthparts called galeae, which, after the moth ecloses, quickly assemble themselves into a straw-like structure. Normally coiled into a tight spiral, the proboscis has muscles that allow it to uncoil and flex in various ways. Darwin's Moth (also known as the Predicta Hawk Moth), *Xanthopan praedicta*, of Madagascar guides its 12-in (300-mm) proboscis into the narrow opening of a star orchid flower (*Angraecum sesquipedale*) while hovering over it like a helicopter. The interplay of hydrophilic (water-loving) and hydrophobic (water-repelling) properties of the proboscis allows it to pick up and transfer liquid from a flower into the moth's oral cavity. This is partially achieved by employing the same forces that make water rise inside a narrow straw and capillary—the narrower the tube, the higher water rises.

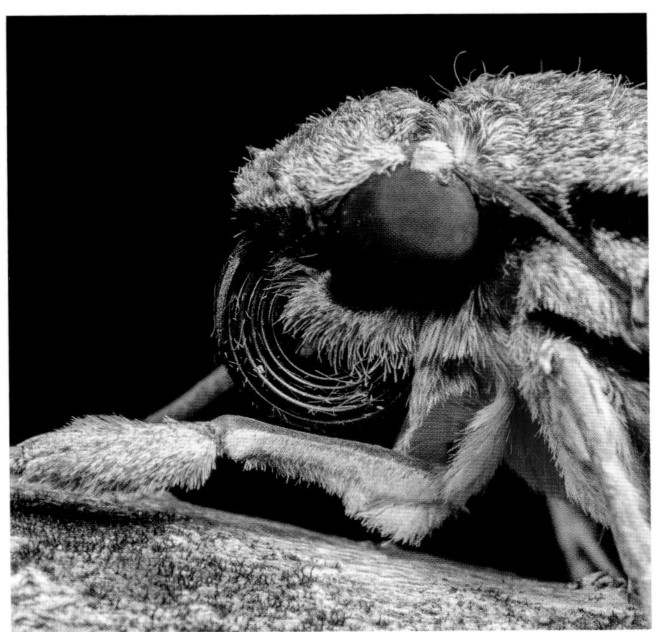

↑ Darwin's Moth (*Xanthopan praedicta*) rests on its perch, its long proboscis tightly coiled when not in use.

of cross-pollination. Instead of pollen, such orchids produce pollinia (pollen packages) that stick to the moth when the proboscis is fully inserted. The transfer of the pollinium to another flower then happens in a similar manner. Orchids, one of the most diverse groups of flowers, frequently employ such deceptions to manipulate different insects.

EXTRAFLORAL NECTAR

Many plants produce extrafloral nectar from nectar-secreting glands outside the flower, sometimes below the flower and at the base of the leaves. While costly to the plant, it can pay dividends: ants, for example, patrol plants such as passion vines (*Passiflora* spp.) and rid them of herbivorous enemies. Extrafloral nectar, however, can also attract "freeloaders," such as the Tropical Sod Webworm Moth (*Herpetogramma phaeopteralis*), which at night may gather around extrafloral nectaries of passion vines. Fertility in the Soybean Looper Moth (*Pseudoplusia includens*) appears to increase when they feed on the extrafloral nectaries of cotton. While this type of insect–plant interaction does not assist in pollination, extrafloral nectaries may direct the less helpful visitors away from the flowers, thus preserving flower nectar for its desired pollinators.

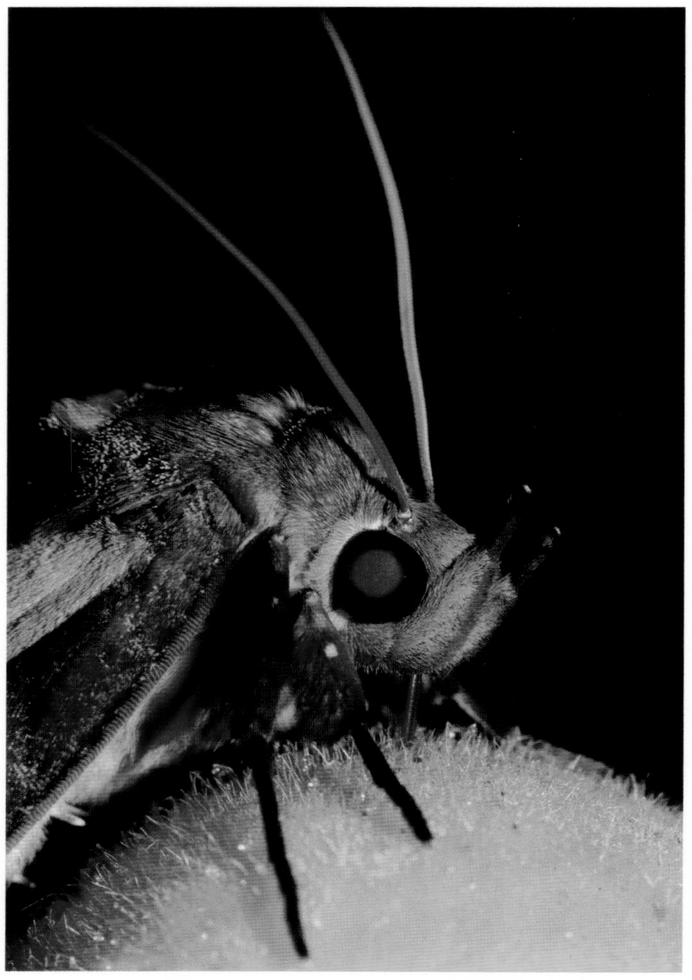

← A fruit-piercing moth in the Calpini tribe of Erebidae feeds at night.

→ *Dysphania militaris.* Diurnal male moths frequently exhibit puddling behavior akin to that of butterflies, collecting minerals from wet ground by sipping and discharging water.

OTHER FOODS

Moths are certainly no strangers to bizarre foods and, like butterflies, frequently drink from other sources. Sugar-rich sources that attract them include tree sap, honey in bee nests, and fruit. Different species of Death's-head Hawk Moths in the genus *Atropos* are infamous for entering beehives, behavior that originated from the species raiding wild bee nests, and drinking honey and nectar. The best way to attract underwing moths (*Catocala*) is by spreading a fermented mixture of sugar, wine, and bananas on tree trunks.

Some moths can be caught in bait traps, using rotten bananas. This is especially true for certain large erebids, such as the Black Witch Moth (*Ascalapha odorata*) of the American tropics, which feeds on fruit that has fallen from orchard and rainforest trees. Like butterflies, many erebid moths can be attracted to baits made of fermented fruit. Some fruit-piercing moths, such as the Common Fruit-piercing Moth (*Eudocima phalonia*), have developed specialized tough and barbed proboscis structures to penetrate the skin of fruit. *Hemiceratoides hieroglyphica* moths of Madagascar visit sleeping birds

and drink their tears (eyes of birds and reptiles produce electrolytes similar to our tears to keep their eyes healthy). Tear-drinking also occurs in many other species elsewhere in the world; when a moth irritates the eye of a sleeping bird with its proboscis, it is able to "harvest" its tears (by day, butterflies are frequently seen gathering around turtles to drink their tears). Salts are the main attractants for moths to tears, but in the tropical *Calyptra* spp., also known as vampire moths, drinking tears evolved into drinking the blood of vertebrate animals. Puddling—collecting salts from riverbanks and puddles on dirt roads—is particularly common in day-flying moths, such as sunset moths (Uraniidae) and many day-flying geometrids. Though this behavior is more commonly observed in butterflies, male moths may also come to other less appetizing baits, such as rotting fish or animals, in search of salts and nitrogen.

Predators and defenses

Moths have devised ingenious defenses to protect against the many threats they face. Birds, mammals, lizards, toads, spiders, and wasps, all seek them out, especially at the larval stage, as caterpillars are an abundant and excellent source of protein. Parasitoids and disease also take a considerable toll. As many as 99 percent of individuals die before they can reproduce. To protect themselves in this hostile world, moths and their caterpillars have had to evolve an arsenal of weapons and strategies such as spines, poisons, sounds, erratic flight patterns, camouflage, and mimicry.

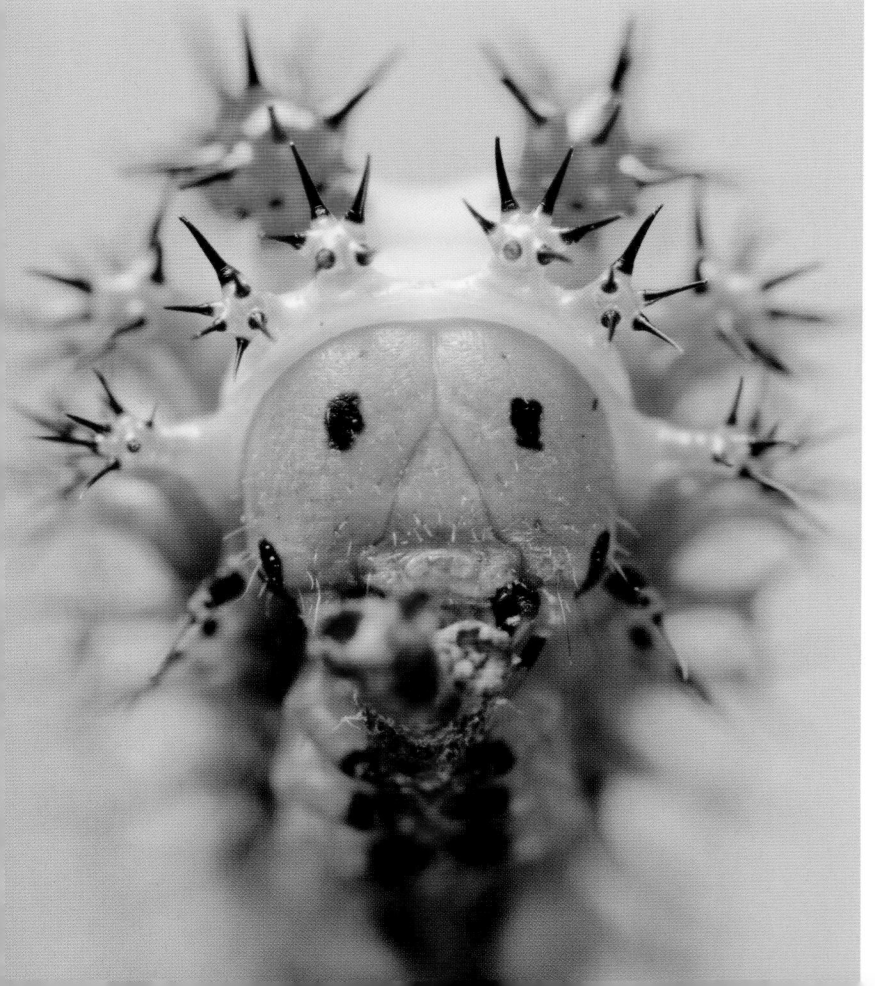

← The caterpillar of the Cecropia Moth (*Hyalophora cecropia*) is both brightly colored and has formidable defensive spines to deter birds from attempting to eat and swallow it.

→ Many tiger moth caterpillars (known as "woolly bears") are equipped with dense hairs that make them less desirable prey and even protect them from some parasitoids.

SPINES, HAIRS, SPURS, AND CROCHETS

Many caterpillars have rigid spines that break off and damage the soft tissues in the mouth of creatures that try to eat them, delivering a longer lasting impact, sometimes backed up by toxins, as in slug moths (Limacodidae). Spines of woolly bears and other Erebidae are more flexible, and can be either short but tough, or long but soft. The latter may easily come off, leaving an attacking predator with a mouthful of hairs that, too, can stick in their soft tissues. Caterpillars of many moths, such as the large silk moths (Saturniidae), defend themselves against vertebrate predators such as lizards by clinging onto branches with their crochets (hooks on prolegs), so they are difficult to tear off. This passive defense strategy is especially effective in species that also have spines, forcing predators to let them go. Others, such as *Spodoptera* noctuid caterpillars, simply let go at the slightest sign of danger, dropping to the ground and out of sight.

Many pupae are protected by a tough cocoon made of silk and sometimes other substances, such as caterpillar hairs and secretions. This, together with a thick pupal case, made of a fibrous substance called chitin, shields pupae from the elements and from predators. Although not common, the pupae of some species, such as the Southern Flannel Moth (*Megalopyge opercularis*), gain additional protection from spines on their sides, and in the Imperial Moth (*Eacles imperialis*), the cremaster, which is normally a set of hooks on the last abdominal segment, evolved into a sharp cutting organ; by moving its abdomen vigorously, the pupa (lacking cocoon protection in this species) can still defend itself against a predator.

The defenses of adult hawk moths (Sphingidae) include sharp tibial spurs and strong flight muscles—a combination that makes it very difficult for a bird or a bat to capture and hold on to them. Strength and size are also an advantage when repelling attackers: the Agrippina Moth (*Thysania agrippina*) and the Atlas Moth (*Attacus atlas*), with wingspans exceeding 1 ft (330 mm), are both powerful and too large for most normal moth predators to capture.

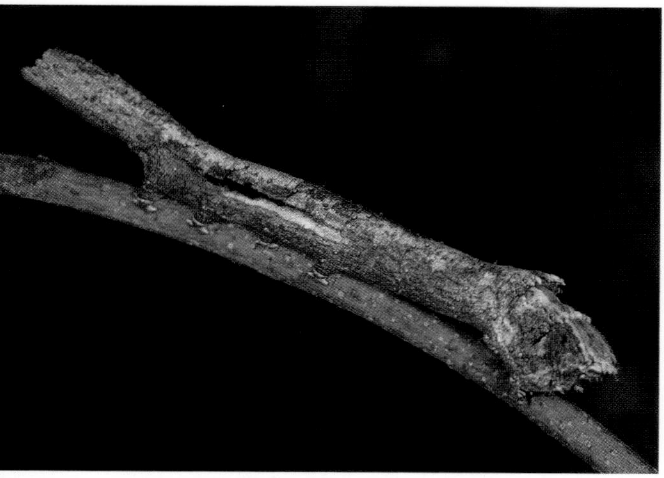

← A caterpillar in Yunnan, China, is easily mistaken for part of the branch it rests on.

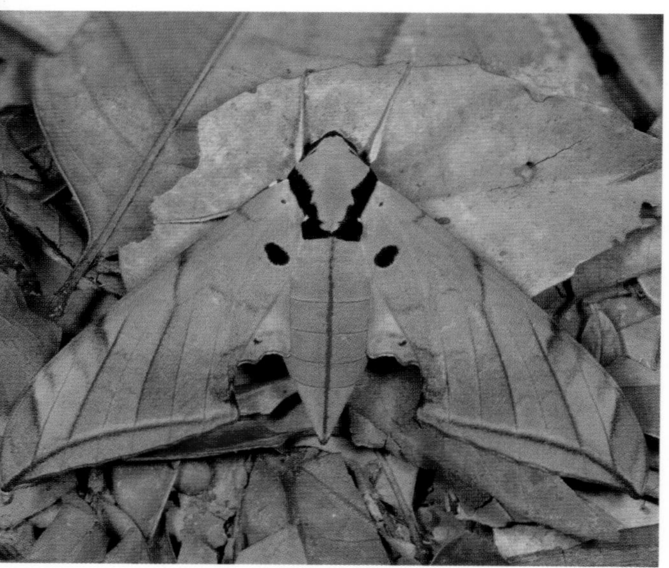

↙ The cryptic leaf-mimicking forewings of the Violet Gliding Hawk Moth (*Ambulyx liturata*) conceal brightly colored hind wings.

MASTERS OF CAMOUFLAGE

While some moths and their caterpillars have vivid aposematic coloring, most species are cryptic—resembling bark, sticks, or leaves to blend in with their surroundings. Many moths can also arrange their bodies and wings in remarkably diverse postures to become almost indistinguishable from the plant part they rest on. Young caterpillars that feed on top of a leaf may look like bird or lizard droppings, and occasionally this is true of adults, too, such as the Frilly Grass Tubeworm Moth (*Acrolophus mycetophagus*). The more common camouflage coloring is green: caterpillars can easily achieve this by simply relying on the color of the food in their gut showing through their translucent skin. Many adult moths of the Geometridae family are green, although green pigment in scales (called geoverdin) is trickier to produce. Perhaps the most ingenious example of cryptic coloration is found in the European Buff-tip moth (*Phalera bucephala*), in which both ends resemble broken-off twigs with birch bark coloration in between.

Phenotypic plasticity—the ability of the same genes to produce different phenotypes depending on environment—may be more widespread in caterpillars and adult moths than is currently known. Lowering pupation temperature, for example, has been shown to produce darker adult moths in some species. Polyphenism is another quite common form of phenotypic plasticity and occurs when seasonal conditions produce different phenotypes. For example, Io Moths (*Automeris io*) in north Florida that develop through into adults without interruption have yellow forewings, whereas the forewings of those that enter diapause, overwinter, and emerge the following summer are orange-brown.

THE FAMOUS PEPPERED MOTH: PAST AND PRESENT

Many moths also resemble lichens on tree bark, and a frequently quoted example from history reveals how quickly species can adapt to selective pressures. In the 1800s, during Britain's Industrial Revolution, air pollution killed off lichen on trees, and within decades in industrial areas, the common lighter form of Peppered Moth (*Biston betularia*), shown here, disappeared because the moths were more visible to bird predators, and a rare black form dominated, providing better camouflage. These variations in the Peppered Moth have been shown to be genetically determined, heritable, and we now finally understand the underlying genetics. A combination of genetic variability and natural selection by birds, which was recently shown around Cambridge, England, to favor the cryptic form by about 10 percent, compared to a less cryptic one, can quickly change the appearance of these moths in any given locality. Caterpillar colors and forms can vary considerably between instars—perhaps resembling a twig at one stage and a leaf at another, and frequently caterpillars, too, like adult moths, have several genetically determined forms to minimize their chances of being wiped out by birds. More remarkably, the Peppered Moth caterpillar also has light-sensitive elements in its cuticle, which enables it to make chameleon-like color adjustments stimulated by changing surroundings.

↑ The Gaudy Sphinx (*Eumorpha labruscae*) caterpillar, found from Argentina to the southeastern United States, is cryptic, but when disturbed, raises and inflates its thorax to appear threatening.

↗ *Hemeroplanes triptolemus* in the jungles around Rio Napo, Peru. This sphinx moth larva is a convincing venomous pit viper snake mimic, despite its small size—apparently, many birds cannot tell the difference.

GENIUSES OF DECEPTION

In Io Moths, an adult's camouflage is also combined with bright coloration. To confuse a bird or mammal predator, this and many other moths might suddenly expose a dramatic eyespot or striped pattern from beneath their leaf-shaped cryptically colored forewings. This ploy works in reverse in underwing moths (*Catocala* spp.); if a bird attack results in a chase, these moths' flashing colors disappear when they land on a tree, so they blend in with the bark, throwing the bird off track. In many toxic species, vivid coloration is their main defense: when discovered, they drop to the ground to expose their bright colors and make no attempt to flee, relying instead on their bitter smell and taste to protect them.

Some moth caterpillars can startle predators by changing their appearance to look like a snake head, either by having eyespots on their posterior segments or

← Snake mimicry as a defense is common among caterpillars, such as this larva of a fruit-piercing moth (*Phyllodes* spp.) in Yunnan, China.

→ The day-flying moth *Mesothen temperata*, found in Costa Rica, is a convincing wasp mimic.

A MOTH THAT LOOKS LIKE LIZARD DROPPINGS

The brown-and-white Frilly Grass Tubeworm (*Acrolophus mycetophagus*) moths in the tubeworm moth family Acrolophidae are easily mistaken for lizard droppings—and frequently rest on leaves where lizard excrement is found. However, close examination of this tiny moth, which occurs throughout the southeastern United States, reveals the intricate detail of its ornate head region, with its wild plumes of hairs that may help camouflage the moth and likely have other, as yet unknown, functions. Though a common moth, its caterpillars have been reared only once—on bracket fungus that attacks the wood of living trees, hence its Latin species name mycetophagus (mushroom-eating). *Acrolophus* sp., shown here, represents a large New World genus of more than 220 species, 63 of which are known in the United States. Their caterpillars develop in the soil, feeding on organic matter, and likely on fungus.

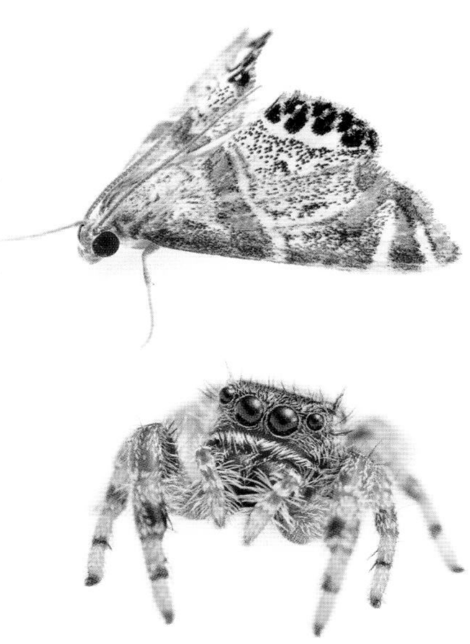

↑ The eyespots and hind wing lines on the *Petrophila jaliscalis* crambid moth mimic the shiny eyes and facial pattern of small jumping spiders, such as *Habronattus* spp.

on their thorax, which they inflate when threatened. In many species, the caterpillar's tail end strongly resembles its head, enabling the larva to escape, or hurt the predator with spines of secretions. For instance, many hawk moth caterpillars have a posterior horn that is rather sharp and spiny. The Puss Moth caterpillar (*Cerura vinula*) can secrete formic acid. When attacked, it squirts this from its rear end, simultaneously extending blood-red posterior filaments.

The bright spots on the wing margin of many moths can help defend them from predators by distracting attention from their head. A bird or a spider that would normally attack the head region may attack the moth's wing fringe instead, which will break away with little harm to the moth, allowing it to escape. Even more deceptive are the remarkable metalmark moths (Choreutidae) in the genus *Brenthia* which, as a recent study in Costa Rica revealed, looks and behaves so much like a jumping spider that, if one of these spiders encounters the moth, instead of hunting it down, it puts on a sexual display.

On occasion, moths may also be involved in Batesian mimicry—when a non-toxic species protects itself by mimicking another common toxic or stinging species. All clearwing moths (Sesiidae), for example, are practically indistinguishable from stinging wasps, and some diurnal hawk moths, such as the Hummingbird Clearwing (*Hemaris thysbe*), mimic bumblebees.

CHEMICAL DEFENSES

Some of the caterpillars that can feed on toxic plants absorb a variety of secondary compounds, varying from acids to alkaloids, which help repel attackers. All these chemicals (which are the plants' own defenses against herbivores) can make caterpillars and the resulting pupae and adults taste foul to predators and may even be poisonous—an effective chemical defense against many of their natural enemies. Some caterpillars have no need to acquire plant toxins as they have glands that enable them to produce their own, and hollow spines through which they can inject their poisons into predators.

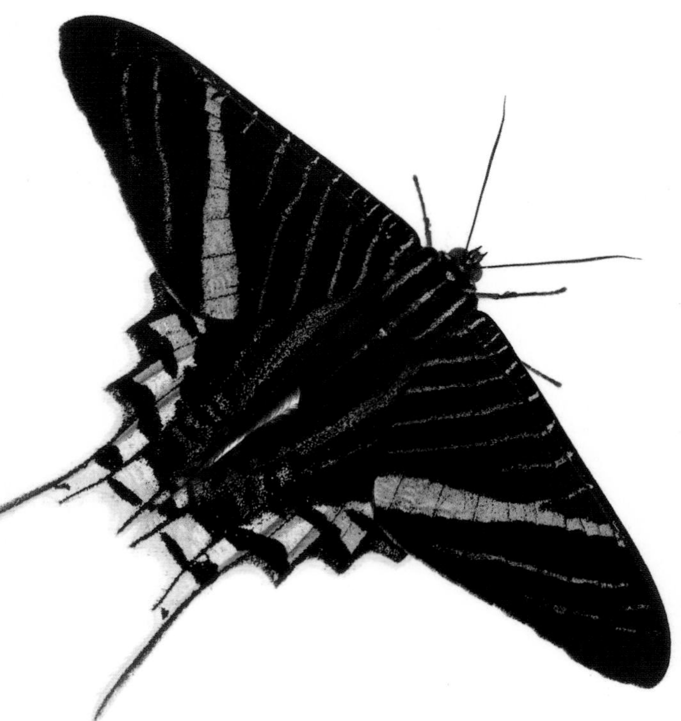

↑ A brightly colored and chemically defended Green-banded Urania Moth (*Urania leilus*).

Where defensive chemicals persist to the adult stage, their protection may have enabled some moth species to switch to a diurnal lifestyle. Rather than relying on the cover of darkness or fast flight to escape, their bitter taste will make any predatory bird release them. These moths often resemble toxic butterflies or beetles—a phenomenon known as Müllerian mimicry—as a similar appearance intensifies the warning message to predators for their mutual benefit. Birds easily memorize the bright aposematic colors of these insects, making their toxic warnings a more effective defense than camouflage, speed, or maneuverability. By night, certain tiger moths, such as the Dogbane Tiger Moth (*Cycnia tenera*) and the Polka-dot Wasp Moth (*Syntomeida epilais*), which are also defended by plant chemicals, work together in a similar way, communicating their distastefulness to predatory bats via similar sound signals.

DODGING THE DANGER

The adult moth's last, but potent defense is its ability to fly away from danger. Hawk moths, often seen zooming between flowers, not only by night, but by day and at dusk, are the fastest species and can outstrip most predators. Large silk moths are much slower, but some have special adaptations, such as tails on their hind wings that make their flight pattern more irregular and unpredictable. Recent studies have shown that these tails can confuse bat predators, which hunt by echolocation (detection via sound waves). Many smaller moths will detect the bat signals and attempt to fly away, but the tailed moths' wing shape also fools them by deflecting attacks from the head region toward the flashy tails.

↖ Diurnal, brightly colored arctiines, such as this Princely Tiger Moth (*Chrysocale principalis*) in the Mexican highlands, are usually chemically defended.

← The Ornate Bella Moth (*Utetheisa ornatrix*) caterpillar feeds on the alkaloid-rich flowers of a rattlebox plant (*Crotalaria*), from which it sequesters defensive compounds.

Parasitoids, parasites, fungi, and pathogens

While moths at all stages have developed numerous chemical, physical, and strategic defenses against larger predators, there are other enemies that present additional challenges. Parasitoids and microbes are much harder for a moth to resist, especially in its early stages, and they take a heavy toll on Lepidoptera life. One key factor that makes them so deadly is that they are often specialists. Just as caterpillars, in the course of evolution, have fine-tuned their ability to find a host, overcome its defenses, and feed on it, so too have these insidious adversaries, which inhabit and feed on caterpillars (and sometimes eggs or pupae), draining them of life. Some of these natural enemies are such effective killers that humans have weaponized them, for instance, raising and releasing parasitoids to combat invasive species of moth, or using bacterial DNA to produce pesticides.

WASP PARASITOIDS

Unlike true parasites, which do not kill their hosts (though they may weaken them enough that they die), parasitoids, such as braconid and ichneumonid wasps, kill the immature moth, consuming it little by little. These wasps lay their eggs inside the soft-bodied moth larvae, using their sharp, pointed ovipositor to penetrate the cuticle. Wasp larvae hatching from the eggs then consume the caterpillar from within. Just like their prey, specialist parasitoids can become impervious to a plant's toxins. For instance, the braconid wasp parasitoid *Cardiochiles nigriceps*, which attacks the Tobacco Budworm (*Chloridea virescens*), has no problem overcoming the nicotine-rich toxic environment of the caterpillar feeding on tobacco.

Like a moth that can pick up pheromones from far away, female parasitoid wasps sense volatile chemicals produced by all tissues of a host plant and by a feeding caterpillar. Some ichneumonid wasps develop long, syringe-like ovipositors (tubular organs that deposit eggs), which they can guide through soil, a hole in wood, or a cocoon to reach caterpillars or pupae in places where they would normally be safe from other predators. Certain caterpillars, however, such as the Subflexus Straw Moth (*Chloridea subflexa*) acquire chemical crypsis to camouflage them against their host plant, which helps hide them from both predators and parasitoids. Because the Subflexus Straw Moth avoids chewing on the leaves, the plant does not produce the "SOS" volatiles that are detected by parasitoids, and it also gains protection from feeding inside the ground cherry's papery husk.

Parasitoids can also locate moths' eggs, and Trichogrammatidae wasps, which are among the smallest insects, ranging from 0.3 mm to 1.2 mm in size, can fully develop inside a single Lepidoptera egg, then hatch from it in place of the neonate caterpillar. These wasps have even developed certain special

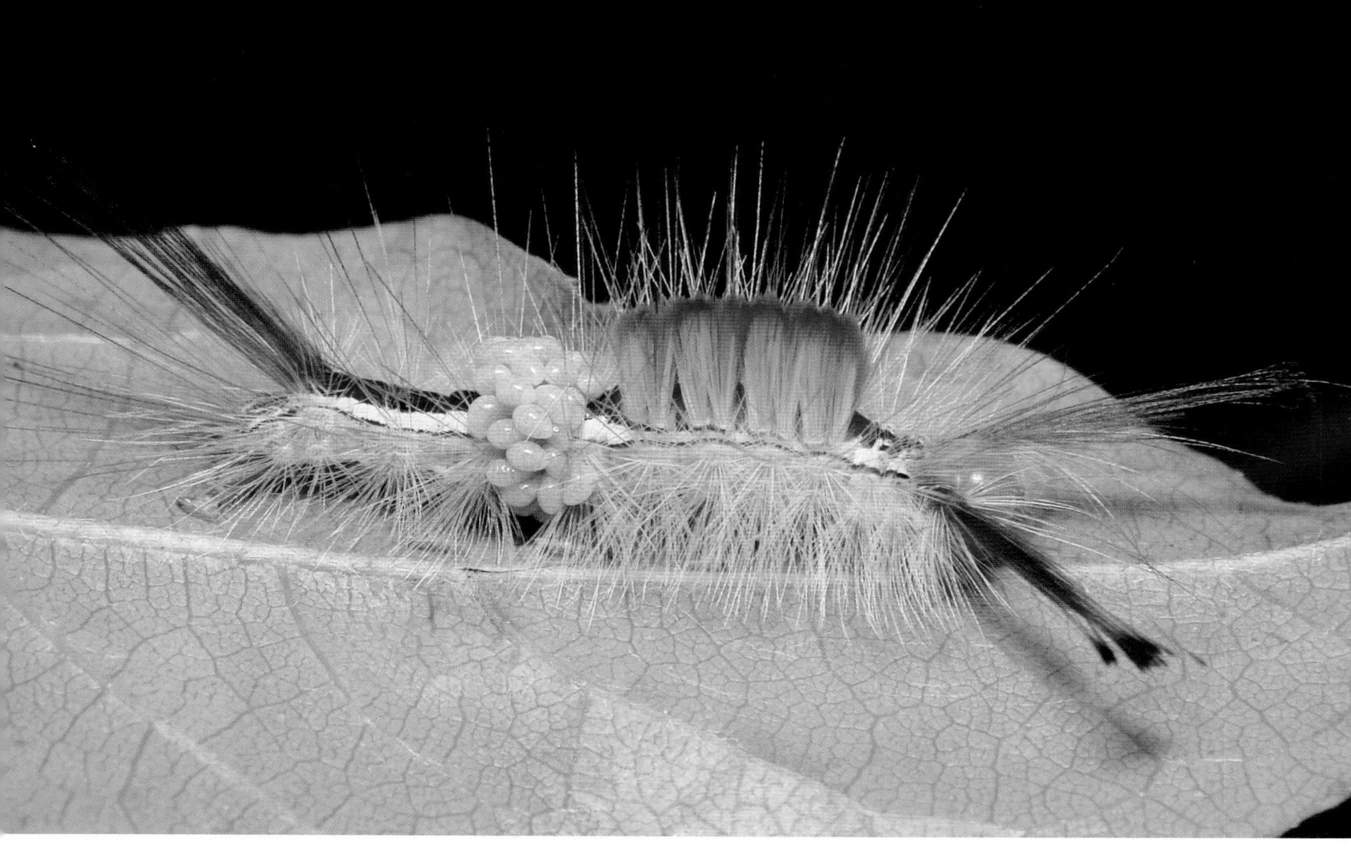

adaptations at a cellular level, such as neurons without a nucleus, in order to undergo such miniaturization.

Caterpillars often hide in shelters made of rolled leaves, coming out only by night to feed, or they may feed only inside stems, leaves, flowers, and fruit, and make dense communal silk nests, as Fall Webworms (*Hyphantria cunea*) and tent caterpillars (*Malacosoma* spp.) do. While the long ovipositors of some ichneumonid wasps can reach the larvae hidden inside such shelters, hiding and night-feeding greatly reduces parasitism. The vast number of progeny that moths produce generally helps maintain a stable population of adults, but this will fluctuate as populations of their natural enemies increase or decrease.

Although parasitoids have an almost infinite capacity to destroy moth larvae, they also need to spare a fair number in order to maintain their own population. If a female wasp parasitoid finds a group of caterpillars, it rarely parasitizes more than 50 percent of them. Chemical markers left by this attacker are also likely to repel a subsequent parasitoid of the same species, so it will move on to look for a new, unparasitized batch. Sometimes, however, different species of parasitoid may attack the same caterpillar, in which case, the consequent larvae must fight it out within the confines of their host's hemolymph—one of many unsavory aspects of insect biology that has doubtless inspired sci-fi movie makers.

↑ Some parasitoid wasps, such as members of the Eulophidae family, feed externally. Their mature larvae resemble a bunch of grapes on the caterpillar's surface and, here, are sucking the lymph of a tussock moth caterpillar.

ATTACK AND DEFENSE

Some wasps paralyze their caterpillars by injecting them with poisons before laying eggs. This prevents a caterpillar from thrashing around and biting the wasp—its first line of defense. Parasitoids also feed on their host's lymphatic fluids while the caterpillar is subdued. However, the mechanical defenses that caterpillars use so effectively against larger predators can help protect them against some, especially generalist, parasitoids. For instance, in one study, a braconid parasitoid, *Meteorus pulchricornis*, was able to parasitize around 90 percent of smooth noctuid larvae but, when offered hairy larvae of a Gypsy Moth (*Lymantria dispar*), its success rate dropped to just over 20 percent, only increasing to more than 90 percent when the caterpillars' hairs were removed.

When the caterpillar's immune system encounters a foreign body, it will frequently, though not always, encapsulate it with hemocytes—cells in insect hemolymph that recognize foreign objects and defend against them, rather as white blood cells operate in the human immune system. While these cells may recognize and encapsulate eggs of the more generalist attackers, some specialist parasitoids have evolved the ability to be invisible to hemocytes.

In two large families of parasitoid wasps—Braconidae and Platygastridae—the parasitoid egg developing inside a caterpillar produces teratocytes. These special cells, which result from the breakup of membrane surrounding the wasp embryo, can affect the level of proteins concentrated in the caterpillar's hemolymph. Together with the venom and particles

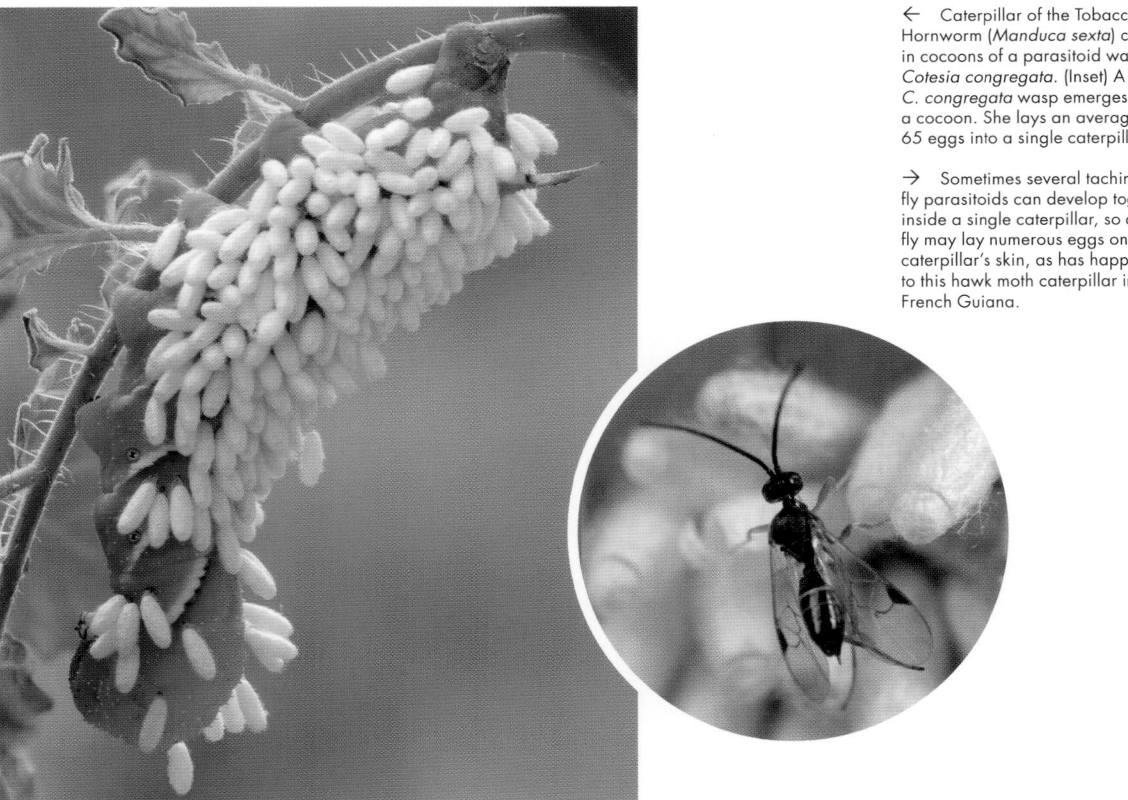

← Caterpillar of the Tobacco Hornworm (*Manduca sexta*) covered in cocoons of a parasitoid wasp, *Cotesia congregata*. (Inset) A female *C. congregata* wasp emerges from a cocoon. She lays an average of 65 eggs into a single caterpillar.

→ Sometimes several tachinid fly parasitoids can develop together inside a single caterpillar, so a female fly may lay numerous eggs on a caterpillar's skin, as has happened to this hawk moth caterpillar in French Guiana.

of a polydnavirus (a special type of insect virus that the female wasp injects with its eggs), they suppress the caterpillar's immune defenses and manipulate its hormonal activity, prolonging its development for the benefit of the parasitoid larvae. Some parasitized caterpillars even live on as "zombies," spreading the cocoons of parasitoids and guarding them after the parasitoid wasps emerge from their host.

FINELY TIMED, FATAL FLY ATTACKS

Flies in the family Tachinidae may also be specialists and generalists, either targeting one or a range of moth species. While hairs and spines can protect a caterpillar from the ovipositors of wasp parasitoids, they may not help it escape tachinid flies. Many tachinid species have developed ovoviviparity, meaning that their eggs are laid in such an advanced stage of embryonic development that young larvae emerge immediately, or even full viviparity where they produce larvae rather than eggs. Frequent molting may help a caterpillar escape being parasitized, but only if it is lucky enough to molt shortly after a tachinid lays its egg. If a fly parasitoid lays eggs on the surface of a caterpillar, the hatchling larvae burrow into the skin to feed inside. In many tachinids, more than one larva can successfully develop inside a single caterpillar, and the size of the consequent flies may vary depending on how many parasitoids shared their host. In others, such as *Chetogena scutellaris*, which attacks caterpillars such as the Cabbage Looper (*Trichoplusia ni*), only one adult fly, and rarely two adult flies, can develop inside one host, no matter how many eggs are laid, though the more eggs laid by the female fly, the higher the chances of successful parasitism.

Fly larvae feed inside their host until the caterpillar goes to pupate. Then they emerge from the pupating caterpillar (or sometimes the moth pupa) and crawl away to pupate, their skin hardening to form a protective, sclerotized puparium. The development of both wasp and fly parasitoids is remarkably fine-tuned to the development of their hosts; parasitoid larvae avoid damaging the inner organs of their host until the very end, so that the host can continue to grow and supply them with food.

Unlike some wasp parasitoids that seek out younger caterpillars with less developed immune systems, tachinid flies appear to attack more mature caterpillars, but may overwhelm their immune system with the number of developing larvae. In some species, they also manipulate the behavior of caterpillars, prompting them to eat more or less.

LYMPH-SUCKING AND WORM PARASITES

Biting midges (Ceratopogonidae), the "no-see-ums," frequently attack humans and suck their blood. While an insect's lymphatic fluid lacks the red blood cells that human blood contains, it remains a protein-rich meal for these tiny flies, specifically *Forcipomyia* spp., which are known to attack butterflies and moths. Interestingly, these midges have adaptations on their claws that help them cling to scales, so it is possible that they have, to some degree, coevolved with Lepidoptera. By night, females feed from both wing veins and the abdomen, penetrating the chitin with sharp mouthparts to extract hemolymph from large moths, such as the Tropical Swallowtail Moth (*Lyssa zampa*). By day, they feed on a variety of Lepidoptera caterpillars.

The nematodes, ubiquitous roundworms, are frequently parasitic, attacking plant roots and animals, especially if they live in the soil. Moth caterpillars are not an exception; in one study, more than 3 percent of mortality in Fall Armyworms (*Spodoptera frugiperda*) was the result of nematode parasitism.

VIRUSES, FUNGI, AND BACTERIA

Some natural enemies of moths are much smaller than even the nematodes and midges, but no less deadly. A caterpillar that turns black and hangs from a tree branch as if liquefied is likely the victim of a baculoviral infection—one of a group of DNA viruses that target insects. When the black fluid leaks from the dead caterpillar, the baculoviral particles spread and may be ingested by other caterpillars munching on leaves nearby. Certain fungal spores, too, accumulate on leaves and, when ingested by caterpillars, grow invisibly inside them. Like caterpillars affected by parasitoids, these caterpillars may appear normal until later in their development, but turn into "mummies," stuffed with fungus, when they are ready

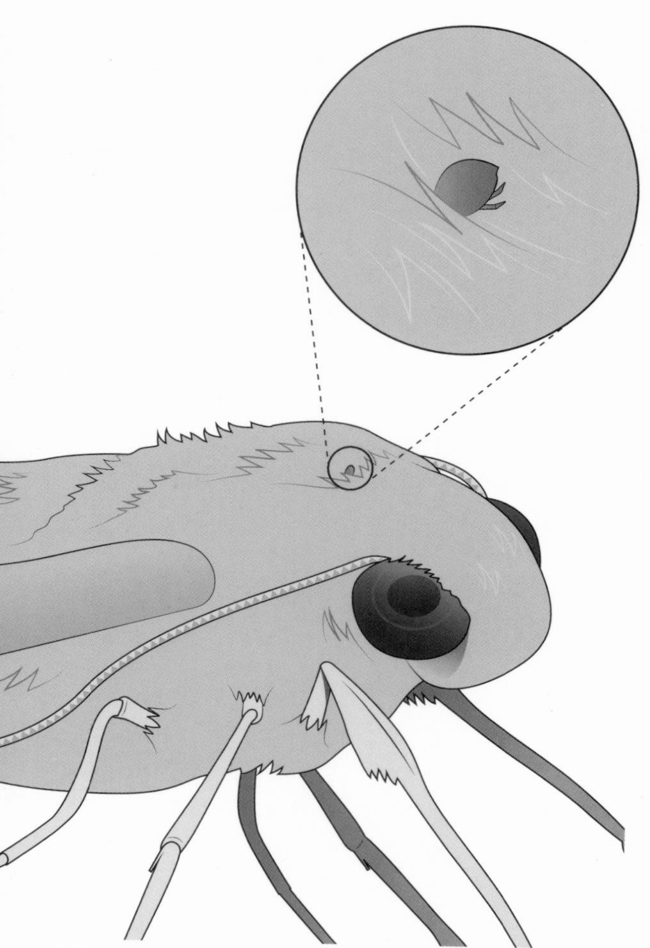

Mite on a sphinx moth
Illustration of a mite feeding on a hawk moth in French Guiana.

to pupate. Occasionally, one can find peculiar-looking caterpillars and adult moth mummies from which fungus called *Cordyceps* has sprouted. These fungi (of which there are over 400 species) produce long stalks, sometimes decorated with sacs containing spores, which adds to their surreal appearance.

Pathogenic bacteria can also kill caterpillars, and to combat them, caterpillar saliva has antibacterial properties. Bacteria are also responsible for one of the most interesting interactions known in the natural world: it occurs between insects, including Lepidoptera, and the microparasite *Wolbachia*. In the early 1900s, researchers noticed that African butterflies in the genus *Acraea* had a sex ratio greatly skewed toward females. It was not until the 1990s, however, that they discovered that the feminization of genetically male individuals is caused by a bacterium, and that an antibiotic treatment can restore the sex ratio to normal. *Wolbachia* were found widely spread among insects, including many moths, and their effect varies. In a recent study, scientists from several Japanese institutions studying a number of Asian Corn Borer Moths (*Ostrinia furnacalis*) that all became female showed that the bacteria hijacked some of the genetic machinery that determines sex expression, leading to the suppression of the masculinizing gene.

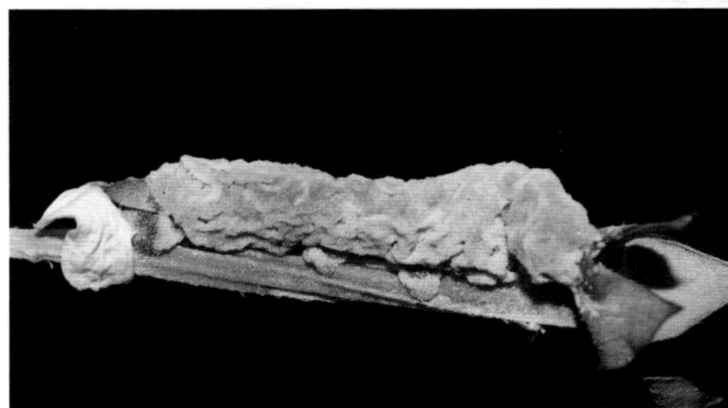

↗ An adult hawk moth in China killed by an entomopathogenic (parasitic on insects) *Akanthomyces* fungus.

→ Caterpillar of the Banded Sphinx (*Eumorpha fasciatus*) killed by entomopathogenic fungus in the family Clavicipitaceae.

Sound production

If butterflies, moths, and their caterpillars seem to be silent insects without a voice, this is largely the result of our selective hearing. The human ear generally detects sounds within a range of 20 Hz to 20 kHz, whereas bats, for example, have a much greater range of 9 kHz to 200 kHz, and they are the kind of audience moths are talking to, as well as to each other. The sounds produced by some butterfly and moth species, such as the nymphalid cracker butterflies (*Hamadryas* spp.) and the Australian whistling moths, are just about audible, but only a close-range, sensitive measurement microphone coupled with an amplifier can pick up most other Lepidoptera sounds, such as chirps, squeaks, clicks, and whistles. These are used variously during courtship, to signal to competitors, and as a defense against predators.

LOVE SONGS

Males of the whistling moths (*Hecatesia* spp.) occupy a territory and produce whistle-clicking sounds in flight and clicking at rest. Like many butterflies, they are territorial and engage in aerial displays with any intruding males. Their bright colors, together with sounds, are part of both territoriality and their courtship ritual. The male of the Small Whistling Moth (*H. exultans*) sits on a leaf and produces "songs" in the form of rapid clicks at intervals of about 5 milliseconds. The clicks may be continuous, but most come in series of 10 to 15 pulses. They are produced by the "castanets," as the paired organs on the moth's specially modified wings were named by the researcher who described them. Males of two other *Hecatesia* species—the Southern Whistling Moth (*H. thyridion*) and the Common Whistling Moth (*H. fenestrata*)—use similar organs to make a clicking-whistling noise, but only in flight.

In the Asian Corn Borer (*Ostrinia furnacalis*), males produce ultrasounds by holding their wings upright and vibrating them. These "songs" can last for several

minutes and, with eight to ten pulses within the 25–100 kHz frequency, they are described as chirps. In one random survey, 70 percent of the species studied produced quiet songs of some sort during courtship. As with pheromones, acoustic signals may be species-specific, helping a male to distinguish an appropriate mate from closely related species. Sound plays an important role in mate location and gaining a prospective mate's acceptance in many species, such as the Polka-dot Wasp Moth (*Syntomeida epilais*) and other day-flying erebid moths.

↑ Female Polka-dot Wasp Moths (*Syntomeida epilais*) emit ultrasounds that enable males to locate them.

HOW MOTHS HEAR

Insects have a hearing organ called a tympanum, which can be located on different body parts. Although not all moths have outer organs to help them hear, they detect sound in much the same way as we do, via membranes that vibrate when a sound wave hits them and excites nerve cells. Such hearing systems can be located anywhere on a moth, including its wings, thorax, abdomen, and even mouthparts. Caterpillars, too, can detect sound waves using hairlike sensory receptors and will respond in different ways, from thrashing or everting tubercles to "freezing" in a camouflaged position, when confronted with sounds that they perceive as dangerous.

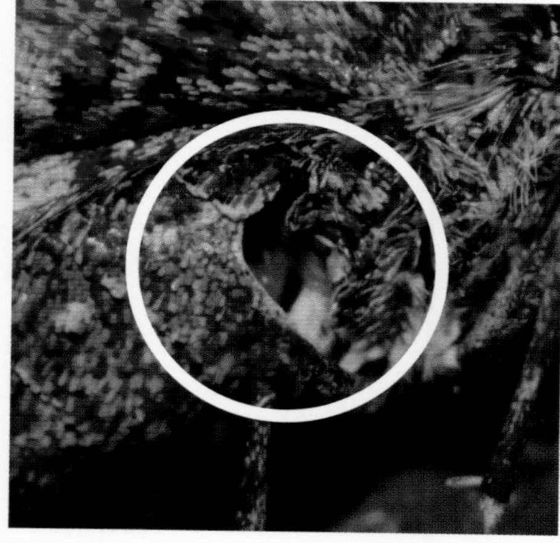

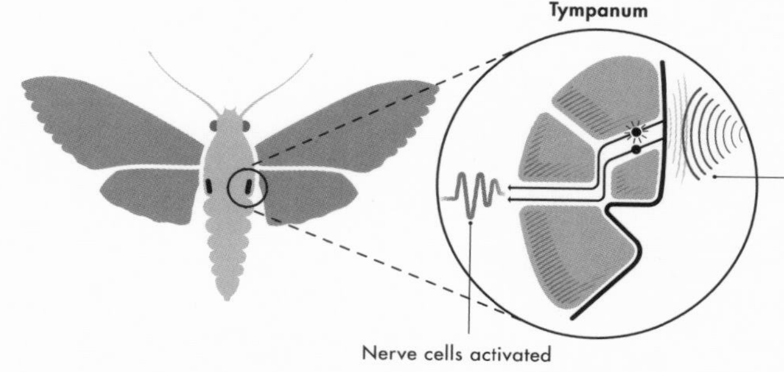

↑ A close-up of a moth ear on the abdomen of the day-flying geometrid moth *Heterusia cruciata*.

WARNING CALLS IN MOTHS AND CATERPILLARS

Metalmark (Riodinidae) butterfly caterpillars communicate with protective ants using sound vibrations created by rubbing the grooves of special rodlike appendages against bumps on their head. No similar system has yet been found in moths, but both caterpillars and larger moths, such as hawk moths and silk moths, emit defensive sounds. Some, such as Walnut Sphinx (*Amorpha juglandis*) caterpillars, produce whistles by contracting their abdomen and letting air escape rapidly from the trachea through the spiracles, while others, such as the Nessus Sphinx caterpillar (*Amphion floridensis*), make a similar sound by exhaling air from their mouth. In a unique adaptation, the adult Death's-head Sphinx Moth (*Acherontia atropos*) expels air from its oral cavity into a shortened proboscis to produce audible squeaky sounds. These caterpillars and moths produce such sounds in response to disturbances, and they help to deter predatory birds.

Caterpillars of the Tobacco Hornworm (*Manduca sexta*), the Polyphemus Moth (*Antheraea polyphemus*), and many other species, have special devices on their mandibles that slide against each other to produce clicking sounds. Sounds produced by owl moth (*Brahmaea* spp.) caterpillars are audible to the human ear. The sound displays can be backed up by real defenses, such as toxins or spikes, or they may be a bluff, designed to startle a predator and fool it into mistaking an edible caterpillar for a dangerous one. Pupae of many moth species can also produce sounds by scraping their abdominal segments against each other while wiggling, but whether these are efficient defenses under any circumstances is unknown.

Hunting bats normally produce ultrasound calls two to ten times per second in the range of 25–100 kHz as they fly around looking for insects to eat, and they increase their cries as they receive an echo and start zooming in on the potential prey. Millions of moths fall prey to bats each night, and, in the course of evolution, have developed a number of defenses. Some larger moths have wing scales that create a metamaterial that can dampen the bats' cries. Others can detect the calls and engage in active defense when their hearing system warns them of the approaching danger. Some moths try to outfly bats, while others, such as tiger moths (Arctiinae, subfamily of Erebidae) can produce clicking ultrasounds via air-filled tymbal organs on the thorax, which have a vibrating membrane at their core. The sounds appear to both interfere with bats' ability to hunt by echolocation and also inform the bats that their prey is potentially distasteful. Some moths create sounds by stridulation—rubbing, for example, a segment of the leg against a stridulatory swelling on the wing, as found in certain Noctuidae species. Hawk moths frequently stridulate by rubbing parts of their genitalia against abdominal segments. Many other types of stridulation mechanisms have evolved in different groups of moths.

↓ A bat pursuing a moth using echolocation.

Moths may produce ultrasound calls to signal that they are distasteful
Bats use ultrasound to locate prey

Moth ultrasound

Some moths make ultrasounds in response to ultrasound cries of a bat used for echolocation. In some moth groups, for example, tiger moths, they warn a bat that they are defended by chemicals and, in the case of hawk moths, by sharp spines.

MOTHS OF TROPICAL RAINFORESTS

Rainforest diversity

While they vary on different continents and at different elevations, the tropical rainforests that span the globe close to the equator are all home to an enormous diversity of plants and animals—including moths. Lepidoptera, predominantly moths, are one of the most diverse orders found in rainforests which, as their name suggests, receive substantial rainfall—up to 400 in (1,016 cm) a year. This, together with a constantly warm temperature that averages around 82°F (28°C), creates their characteristically lush vegetation and towering trees—perfect conditions for abundant wildlife.

DAZZLING MOTHS

Some of the largest and most spectacular moths are found here, together with the greatest diversity of tiny leaf-mining moths. Creatures of the rainforest sport the gaudiest colors, the most unpredictable wing shapes, and the most unusual lifestyles, and moths and their caterpillars are no exception.

↖ The world's rainforests are vanishing at an alarming rate, now covering just half the area they did a century ago. Some scientists predict they will disappear altogether in the next 100 years.

↗ This beautiful geometrid moth, *Eois* spp., was photographed in a cloud rainforest of Ecuador.

Because rainforests also abound in birds, bats, and other moth predators and parasites, they are a challenging environment for Lepidoptera, stretching their ability to adapt and fight for survival. The rich diversity of plants, however, enables each species to find a unique niche. Sadly, just as researchers are beginning to understand these ecosystems, many species are disappearing as a result of deforestation before they have been properly described. Scientists are also noting a dramatic decline in some moths due to climate change—all troubling signs that humanity must change its collective behavior.

SPECTACULAR SPECIES—LARGE AND SMALL

The lofty canopies of tropical forests in Central and South America are home to dramatic Lepidoptera such as the Agrippina Moth (*Thysania agrippina*), also known as the White Witch, with a wingspan of almost 12 in (300 mm). The Atlas Moth (*Attacus atlas*) of Southeast Asia and its relative, the Hercules Moth (*Coscinocera hercules*) of Australia and New Guinea are both the size of a dinner plate. Hercules Moth females have the largest wing surface area of any Lepidoptera, while the hind wings of the males taper into long thin tails, as do those of the Comet Moth (*Argema mittrei*) of Madagascar, which can be up to 6 in (150 mm) in length.

Rainforest ecosystems are also rich in small moths. Their larvae, which often feed by tunneling inside a single leaf, hence the name "leaf miners," are barely visible to the naked eye. While many people find tiny insects less fascinating than giant species, being so small is perhaps a more impressive evolutionary feat, as all the organism's functions, from feeding as a caterpillar and producing hormones, to flying and mating, must be performed within much smaller bodies. Many of these species are spectacular in their own right, with colorful, though tiny wings. As with large moths, the bright colors are thought to help them communicate with each other and also to avoid smaller predators, such as jumping spiders, that abound in rainforest vegetation. There is a very large undescribed diversity of leaf miners in all ecosystems, especially in tropical forests, which harbor representatives of many primitive families no longer found elsewhere.

ABUNDANT RESOURCES TO FUEL DIVERSITY

The quantity of sun and water that rainforests receive encourages fast plant growth, and consequent recycling of huge amounts of biomass (the total mass of organisms within an area) is much greater here compared to other ecosystems. Moths themselves play an important role in the circulation of nutrients in the rainforest ecosystem.

← The Hercules Moth (*Coscinocera hercules*) is one of the world's largest, with a total wing surface area of 46 sq in (300 cm²).

→ Larvae of the Atlas Moth (*Attacus atlas*) can be over 4¼ in (110 mm) long.

Leaves fall to the forest floor and decompose, feeding the soil with nutrients that plants take up through their roots to support their growth. Herbivores speed up the process of leaf decay. It has been estimated that they destroy from 10 to 45 percent of leaf surfaces, depending on the species of tree, its height, and the season, and caterpillars are among the major contributors to this process. They remove some of the nutrients for their own consumption and release the rest as frass (dung) that falls from the trees with the pitter-patter sound of raindrops.

This constant renewal may explain why life is so diverse here: for example, while the UK has some 7 million acres (2.8 million ha) of temperate woodland, it has only around 70 native tree species. By contrast, researchers have identified more than 700 tree species in a single acre of the Amazon rainforest. Similarly, every moth survey conducted within rainforests around the world has concluded that their local diversity is much greater than in other habitats, and samples are likely to include a high proportion of rare species. Studies also show that different moth species are found at different altitudes, where the range of flora may differ, too. Where caterpillars of the same species live at different elevation levels, some may switch to new host plants, which can lead to the development of new moth species, further increasing diversity.

In a decades-long study of moth caterpillars in Costa Rican rainforest, which has already produced dozens of scientific papers, local researchers have collected tens of thousands of caterpillars and reared them through to adult moths. This long-term research has shown that even within the same forest, species that look almost identical as adults can look slightly different as caterpillars, may feed on different host plants, and have different parasitoids eating them from inside. While the naming and describing of new species is a lengthy process, these studies further indicate that rainforests are even more diverse than was previously thought.

Rainforest plants and flowers

The tallest rainforest trees and their canopies create a cathedral-like environment of cool and dark understory where smaller trees and shrubs flourish with microhabitats at different levels. While the forest floor may have little sunlight, fallen trees create light gaps enabling plants to flower.

ESSENTIAL PLANT COMPOUNDS

Many chemically defended moth species feed on lianas—woody vines, such as the elephant creeper (*Entada phaseoloides*) of Africa, Australia, Asia, and the Western Pacific, and *Philodendron* spp. in Latin America—that extend more than 1,000 ft (300 m), coiling upward through the vegetation to the canopy. Caterpillars of colorful diurnal moths, such as *Mapeta xanthomelas* (Pyralidae) and some of the representatives of the Dioptinae subfamily (Notodontidae), prefer pipe vines (*Aristolochia* spp.) and passion vines (*Passiflora* spp.), whose protective acids and alkaloids the moths use for their own defenses. For these and other tropical moths, the secondary plant compounds are crucially important: in the course of evolution, the caterpillars have had to overcome the plants' defenses in order to feed on them and absorb their chemicals. The caterpillars' choices help dictate which host plants the adult female will lay her eggs on, as eggs are frequently positioned where the hatchlings can quickly start feeding. As moths are also more likely to find their mates around their host plants, the distribution of host plants can help define the distribution of a moth species.

The chemicals that moths obtain from their host plants are essential for certain biochemical changes that assist in reproduction. For example, many tropical toxic tiger moths derive chemicals necessary for developing courtship signaling pheromones from the same toxic secondary plant compounds that defend them against the many rainforest predators such as bats and birds.

Some of these tiger moths, including the gaudy *Euchromia* and *Amerila* species in the Solomon Islands, also derive these compounds from wilting plant tissues of the tree heliotrope (*Heliotropium arboretum*), with similar behaviors found elsewhere in the tropics.

FEEDING ON FLOWERS

In Costa Rican rainforests, a giant day-flying butterfly moth, the Sugar Cane Borer Moth (*Amauta cacica*, Castniidae), together with hummingbirds, take nectar from beautiful *Heliconia pogonantha* flowers. Caterpillars of the Sugar Cane Borer Moth tunnel into the roots of *Heliconia*, where they feed and pupate. They are also occasional pests on bananas and plantains and may feed on other members of their Zingiberales order.

Some rainforest moths feed as caterpillars on and take nectar from flowering epiphytes (non-parasitic plants that use other plants to support them), such as orchids and bromeliads, most of which are found in tropical rainforests. In South America, caterpillars of *Castnia therapon* and several other large day-flying castniid moths feed inside the rhizomes and bulbs of orchids, and in Java, several orchids such as *Dendrobium* and *Phalaenopsis* species are host plants for a tuft moth *Urbona chlorocrota* (Nolidae). Hawk moths visit and pollinate orchid flowers, such as the star orchid (*Angraecum sesquipedale*) of Madagascan rainforests. Like orchids, bromeliads also grow on tree trunks at different levels of the forest. In Mexican tropical forest, flowers of *Tillandsia heterophylla* are visited by various noctuid moths and by bats.

↑ The clear wings of this wasp-mimicking tiger moth in Yunnan, China are almost transparent because they lack the scales typical of most Lepidoptera.

← A mating pair of *Euchromia polymena* arctiine moths—a species found throughout Southeast Asia. The moths' bright reds, blues, and yellows are an indication of their chemical defenses.

Coevolution and mimicry

The bright colors and eyespots of many rainforest caterpillars and moths are all part of their defenses. Like warning street signs, reds and yellows, usually combined with black dots or stripes, notify predators that the species they are attacking contain toxic chemicals, while eyespots may confuse an attacker.

THE IMITATION GAME

Among caterpillars, snake mimics are common, as numerous snake species inhabit tropical forests, and are creatures that a bird will seek to avoid. It was on Amazonian expeditions in the mid-nineteenth century that British naturalist Henry Walter Bates (1825–92) conceived his ideas on what later became known as Batesian mimicry—where one harmless species imitates a more dangerous one as a form of defense—something seen in butterflies, which Bates used as an example, and also in moths.

← A slug caterpillar of a cup moth (*Squamosa* spp.), such as this brightly colored example from Yunnan, China, can deliver a venomous sting via its spiny hairs.

Bates produced beautiful illustrated plates of wasplike Brazilian moths in his field journals, published only in 2020, but at the time he did not realize that the moths were toxic too. Around the same time, German naturalist, Johann Friedrich Theodor Müller (1821–97) developed his ideas about well-defended species that imitate each other to emphasize their toxicity to predators—a phenomenon called Müllerian mimicry. The wasplike tiger moths that Bates illustrated and the butterfly-like burnet moths of Southeast Asia that fly by day and mimic aposematic, large swallowtail butterflies are examples of such mimicry. Other diurnal tiger moths, such as *Correbidia* and *Lycomorpha*, so closely resemble the net-winged beetles of the family Lycidae that it is challenging to tell them apart. Both moths and beetles are highly distasteful, so this too is a case of Müllerian mimicry.

NATURAL DIVERSION

It is no accident that both pioneers of tropical biology drew their inspiration from working in Brazil, where they could observe firsthand the biodiversity and the intensity of natural selection, which had earlier fascinated British naturalist Charles Darwin (1809–82). In most mimicry cases, it was the drive for survival that led to the evolution of insects that copy unrelated species to avoid predation by birds, but this strategy can sometimes take an unexpected turn. In the Peruvian Amazon, the hatchlings of one bird species—the cinereous mourner (*Laniocera hypopyrra*)—mimic toxic caterpillars of flannel moths (Megalopygidae) as a protection. Their plumage resembles the hairy flame-colored larvae, and the little birds also move their heads slowly from side to side, just as the hairy caterpillars do.

ART OF TROPICAL MOTHS

In the nineteenth century, the work of Müller, Bates, and Darwin among others, drew the attention of other Europeans to the diversity of tropical insects, including Lepidoptera. Even earlier, at the end of the seventeenth century, the great German-born artist and naturalist, Maria Sibylla Merian (1647–1717) encouraged interest in moth fauna with her beautiful illustrations of representative species, including caterpillars and cocoons, from Europe. In 1705, she produced her *Metamorphosis Insectorum Surinamensium* after traveling to Dutch Suriname. Merian depicted extraordinary South American examples of caterpillars and adult moths, such as the Giant Sphinx Moth (*Cocytius antaeus*) and the Agrippina Moth (*Thysania agrippina*), making many in Europe aware of the riches that lie in the tropical rainforest for the first time. She was also fascinated by metamorphosis, which was poorly understood in her day, and documented it for many species including the tropical ones from Suriname, thus becoming a true pioneer of tropical Lepidoptera research. Moths have since fascinated a number of artists, including Salvador Dalí (1904–89), who incorporated moth imagery in many of his surrealistic paintings. Joseph Scheer continues to create spectacular works in New York, using high-resolution scanners and large-format printers to make giant portraits of pinned moths, including some tropical species.

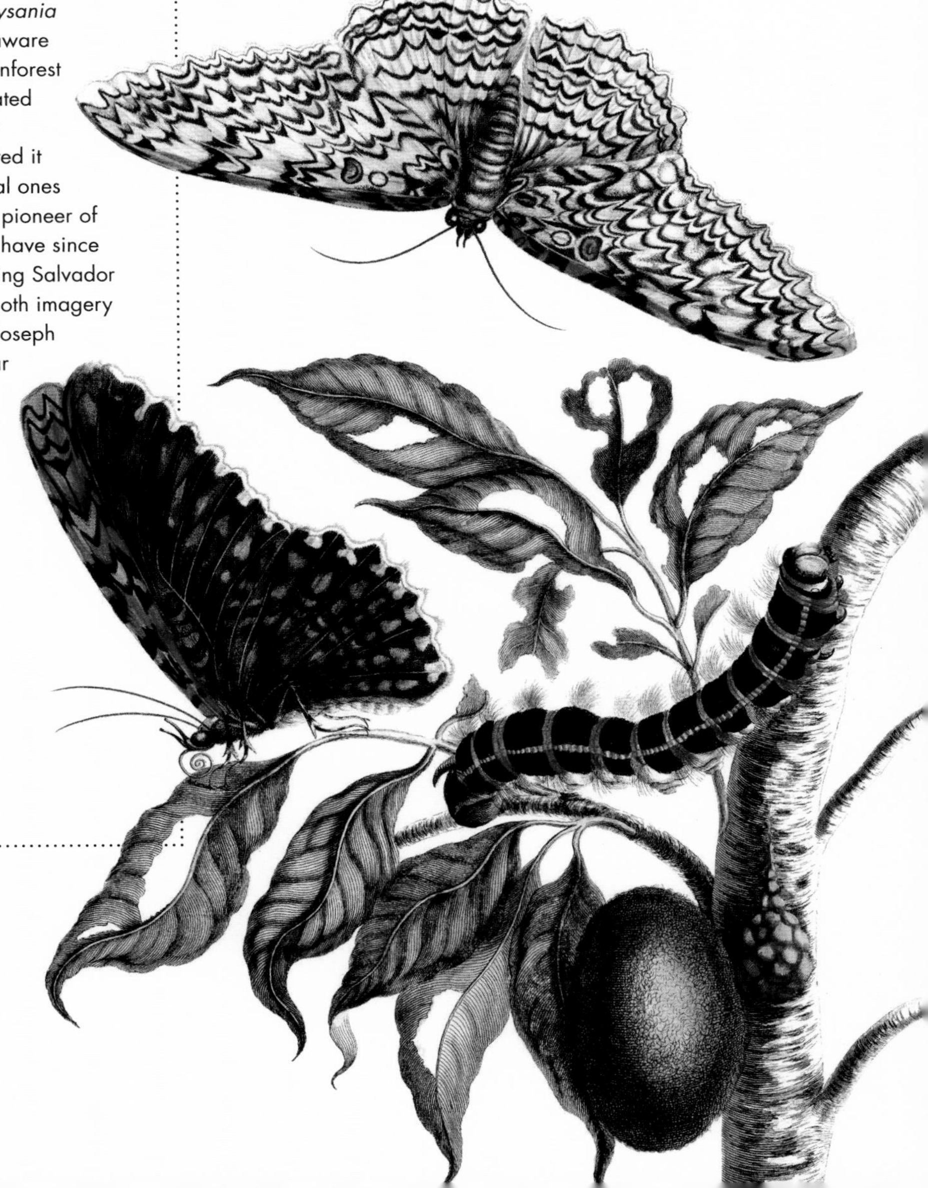

→ The Agrippina Moth (*Thysania agrippina*), which has the largest wingspan of all Lepidoptera, was illustrated in *Metamorphosis Insectorum Surinamensium* by Maria Sybilla Merian in 1705.

Tropical moths with unusual habits

While most moth species have herbivorous caterpillars, some are distinctive for their odd diet at the larval stage and bizarre choice of habitat. They include species living on fungal waste from leafcutter ant colonies and several that are happily cohabiting with sloths, eating these animals' dung.

MOTHS ON ANTS' FUNGAL DETRITUS

Although Thomas de Grey, 6th Baron Walsingham (1843–1919), first described the tubeworm moth *Amydria anceps* in 1914, its biology remained largely unknown until 2003 when a detailed study of its larvae revealed that it feeds exclusively on discarded fungus grown by colonies of leaf-cutting ants of the genus *Atta*. These mostly tropical ants are known from a diverse range of habitats including deserts and tropical forests. Leaving their extensive nests, the ants go foraging for leaf material on which they grow the fungus in deep underground chambers. Once they have fed on the growing fungus, they dump it outside. There, safe from the ants, groups of *A. anceps* larvae feed on the nutritious leftovers, which is mixed with ant excrement. The long, grub-like larvae construct tubes around themselves from silk and debris and later pupate in them. The small adults, variably patterned in shades of brown, with a wingspan of 3⁄16–7⁄16 in (4.5–11 mm), tend to eclose a few days after heavy rains.

MOTHS THAT LIVE IN SLOTH HAIR

In the early twentieth century, two biologists studying moths in the Central and South American rainforest independently described species that live in the fur of sloths. In 1906, German entomologist Arnold Spuler (1869–1937) named his newly discovered Brazilian

↗ Tiny *Cryptoses* moths living in sloths' fur gain protection, and may encourage the growth of algae that the sloths enjoy.

← Leaf-cutter ants feed on fungus that grows on cut leaves. The fungal by-product of their feeding provides food for the tubeworm moth *Amydria anceps*.

moth *Bradypodicola hahneli*, while in 1908, American entomologist Harrison Gray Dyar Jr. (1866–1929) christened his new find *Cryptoses choloepi*. The two moth genera, in the snout moth family Pyralidae, proved to be quite distinct.

Adults of the less studied *B. hahneli* species are believed to lose most of their wings (and hence their capacity to fly) as they burrow deep into the animal's fur, while *Cryptoses* adults can be seen flying up from the fur, sometimes in large numbers. A detailed 2013 study of these species' association with the three-toed sloth concludes that the moths, sloths, and algae in the sloth fur have a complex and highly unusual, mutually beneficial relationship. Despite the danger of predator attacks, three-toed sloths descend from trees to defecate, enabling female *Cryptoses* moths to lay eggs on their droppings. Their caterpillars feed and develop on the excrement, then the adults fly back to the sloths and mate within the security of their fur. The sloths benefit because the presence of the *Cryptoses* moths appears to promote the growth of nutritious green algae in their fur, which they consume while self-grooming and which also contributes to their camouflage in the trees. Three other moth species—*C. waagei*, *C. rufipictus*, and *Bradypophila garbei* are now known to have similar associations with sloths.

Seasonal changes and migrations

Tropical rainforests on different continents experience varying seasonal changes, but all are much more seasonal than was traditionally thought. While all rainforests are evergreen and conditions are relatively mild, the sharp contrast in rainfall between rainy seasons and dry seasons leads to changes in fauna and, sometimes, to distinctive behavioral adaptations.

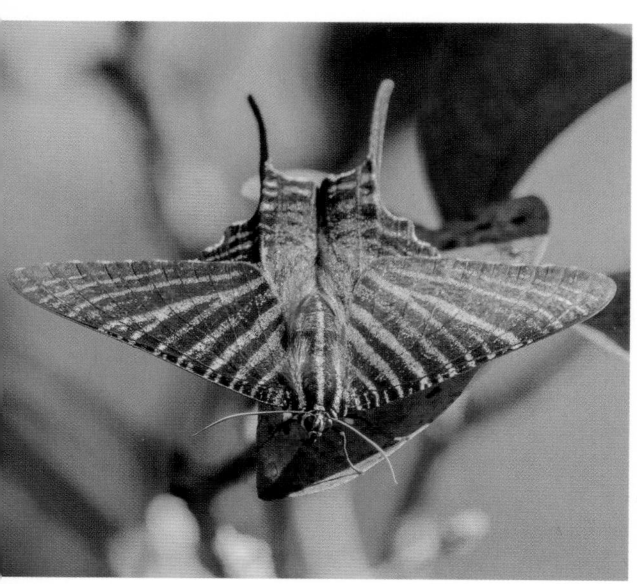

↑ Boisduval's Urania (*Urania boisduvalii*) occurs throughout Cuba; these moths migrate in groups, flying along Cuba's extensive coastline in search of their *Omphalea* host plants.

MIGRATION TRIGGERS

Sunset moths (Uraniidae)—a tropical family of around 700 frequently large and brightly colored species—may fly away as adults in a spectacular mass migration. This happens, for example, in both *Urania* in Latin America and *Chrysiridia* in Madagascar. The migration of the striking Green Page Moth (*Urania fulgens*)—mostly females carrying mature eggs—has been linked to the dry season in Costa Rica, suggesting that it results from the seasonal availability of the species' host plant for the caterpillars to feed on. Increased leaf toxicity may also trigger a migration as, in response to intense caterpillar feeding, the plant produces such concentrated defensive chemicals that the larvae die, forcing the adults to move on to new host plant areas. Both the Green Page Moth and the beautiful Madagascan Sunset Moth (*Chrysiridia rhipheus*) feed on toxic Euphorbiaceae host plants in the genus *Omphalea* and are known to completely defoliate them.

Huge numbers of another spectacular uraniid, the Tropical Swallowtail Moth (*Lyssa zampa*), can occur seasonally by day and by night in Southeast Asia. These moths' caterpillars also feed on Euphorbiaceae, such as the rubber tree and several others. In 2014, in Singapore, Thailand, and Malaysia, researchers reported the mass emergence and later migration of this species, estimating that in June that year between 15,000 and

20,000 mature caterpillars were defoliating an average *Endospermum* tree. It was hardly surprising that the number of citizen scientists' reports of the moth sightings for that year was 50 times greater than the usual annual average.

HAWK MOTH SEASONAL DIVERSITY

The Amazonian rainforest is believed to harbor 80 percent of Brazil's hawk moth species, but just a quarter of them are found exclusively there. Elsewhere, most hawk moths will migrate in order to avoid any temporary seasonal decline in suitable host-plant leaves for their caterpillars to feed on. They can fly long distances and rarely land, even feeding while in flight, and therefore greatly depend on the availability of fuel in the form of nectar. Trees in the genus *Inga* play a key role in hawk moth biology; in the Atlantic rainforest, 70 percent of hawk moth adults feed on their flowers. As a result, hawk moth seasonal diversity in that region correlates with the flowering season of the three *Inga* species. Hawk moths are among the best studied moths due to their large size, and while many local species are threatened and are going extinct, scientists are still discovering new species: for example, several new South American species in the genus *Xylophanes* were described in 2021.

↖ The Tropical Swallowtail Moth (*Lyssa zampa*) migrates out of rainforests in Southeast Asia and is frequently found around lights, even in major cities.

↗ In both the New World and Old World, toxic *Omphalea* spp. serve as the host plants of several sunset moth (Uraniidae) species.

→ Many hawk moths in the Neotropics migrate seasonally in and out of the rainforest in search of host plants and nectar.

In Madagascar, the flight of Darwin's Moth (*Xanthopan praedicta*) that pollinates one of the most beautiful, and now extremely rare, flowers—the star orchid—coincides with the time the flowers open. Only this hawk moth's proboscis—the longest of any moth species—can reach the nectar inside. While this particular plant–pollinator association is known largely because Darwin predicted its existence and because of the star orchid's surreal beauty, such insect–flower relationships, including those that involve moths, are common in rainforests, but most are as yet undescribed.

Effect of deforestation and climate change

Deforestation is a major threat to moth fauna on the planet and especially so in the tropics, where forests are being logged and burned at a staggering rate. In 2019, the World Wildlife Fund reported that the tropics lost 30 soccer fields' worth of trees every minute. Sometimes, there is disturbed habitat left behind after a forest is logged, but most of the diversity of lush rainforest is lost. Frequently, the logged areas are then burned to clear land and to provide additional nutrients for the grass that feeds cattle. Even if these lands are abandoned, as frequently happens when soil is depleted of nutrients, it is much more difficult for plant species to repopulate as rain soon washes away the thin layer of topsoil, exposing rocks.

↓ Deforestation and erosion along the mountains of Serra do Mar in southeast Brazil.

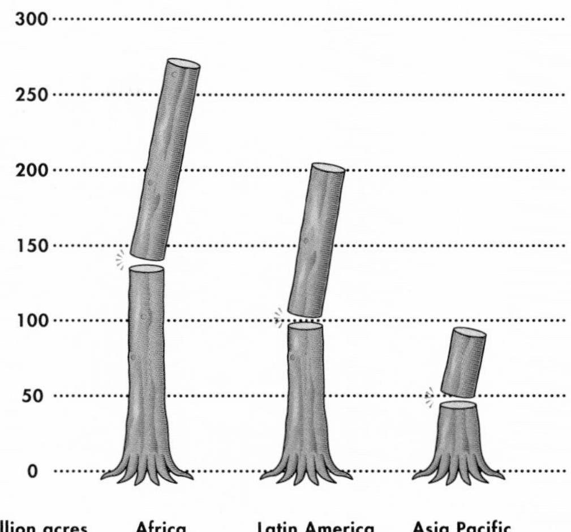

Projected deforestation

Projected tropical deforestation, by region, between 2010 and 2050 (Source WWF, *Living Forests Report* 2011). The most irreversible and shortsighted of all human activities is the ongoing cutting down of tropical rainforests and the loss of their unique wildlife, which includes tens of thousands of unique moth species.

RAINFOREST DESTRUCTION

Tropical islands are where many unique species of moths are found. Madagascar, where most of the moth species are endemic, lost 50 percent of its forests between 1950 and 1985, and the trend has continued since. In the Philippines, the archipelago of 7,641 islands that also harbors many endemic species, most of the lowland forest has been destroyed, and the remnants of rainforest that can be found only in the mountains on some of the islands. Financial help for the poorer societies living in these areas and better education that emphasizes the beauty and importance of biodiversity could help save the remaining rainforest, especially if combined with fast and decisive conservation efforts. But such changes need to happen fast.

On the mainland of South America, the Atlantic Forest (Mata Atlântica), stretches from northeast Brazil along the coast and inland in Argentina and Paraguay. Once, it covered more than 390,000 sq miles (over 1 million sq km), but much of it is now deforested. Studies in the remaining forest, which is still highly vulnerable to logging and clearing for agriculture, have revealed the species richness that still remains, including more than 1,200 tiger moth species—60 percent of all known Brazilian tiger moths and 20 percent of the tiger moths found in Latin America.

Despite deforestation, this area is second only to the Amazon in its biodiversity and new species are still being discovered. The small Serra Bonita mountain nature reserve, near the town of Camacan in Bahia State is one small part of it and home to more than 10,000 moth species—about as many as are found in the entire United States. Here, two beautiful emerald-green notodontids *Chlorosema lemmerae* and *Rosema veachi*, were first described in 2017. Both species were named after people who helped conserve their habitat, which now consists of 6,178 acres (2,500 ha) of privately protected rainforest. With governments around the globe largely failing to protect the environment, such private initiatives offer glimmers of hope.

A CHANGING CLIMATE

Climate change is also a threat, yet even harder to tackle without international governmental initiatives. A 2014 study of tiger moths in Brazil estimated that the likely effects of climate change are so potentially damaging that certain vulnerable species may become extinct. One 50-year-long survey of moth communities in Costa Rica strongly suggests a reduction in moth numbers even in places where the forest cover is increasing as a result of the creation of new conservation areas. While the precise reasons for the decline are not completely understood, these troubling changes appear to strongly correlate with changes in the rainfall and extremely high temperatures, and seem to have accelerated since 2005. As the survival of moth species depends on the seasonal appearance of fresh leaves on the host plants their caterpillars feed on and the timely flowering of plants for nectaring, climate change could have devastating effects.

HYPOCRITA REEDIA

Hypocrita reedia

Iridescent tiger moth

SCIENTIFIC NAME	*Hypocrita reedia* (Schaus, 1910)
FAMILY	Erebidae
NOTABLE FEATURES	Iridescent wings, toxic discharge when disturbed
WINGSPAN	1⅕ in (30 mm)
SIMILAR SPECIES	Several *Hypocrita* spp., such as *H. albimaculata*, *H. arcaei*, and *H. drucei*

This beautiful, iridescent, day-flying tiger moth, best known from Costa Rica, is easily confused with a butterfly as it flies along forest edges and in the understory, its females moving constantly between plants, testing them for suitability for oviposition. This and other *Hypocrita* species, together with all members of their subfamily Pericopinae, are toxic at every life stage because their caterpillars sequester alkaloids from their host plants.

DEFENDED AGAINST PREDATORS

Hypocrita reedia adults produce an abundant, defensive, foaming discharge if caught by a bird, which will usually release the moth and subsequently avoid the species despite its deliberately slow flight and habit of perching on the tops of leaves. The moth's bright iridescent colors, with additional red spots and white stripes, are there to remind predators of any unpleasant earlier encounter. The alkaloids that the moths discharge not only smell and taste foul, but if ingested can cause sickness. The caterpillars of *Hypocrita* are aposematically colored and are protected from predators both chemically and by their long, irritating hairs.

MOTH AND BUTTERFLY IMITATORS

Other moths and some butterflies—both toxic and palatable—mimic the bright coloring and patterning of *Hypocrita* species as a defense against predators. Throughout the Neotropics, there are nearly 40 spectacular *Hypocrita* species, several with much the same wing pattern as that of *H. reedia*.

→ A *Hypocrita reedia* tiger moth in Costa Rica rests conspicuously on a leaf in broad daylight, protected by the bright coloring that signals its toxicity.

EREBUS EPHESPERIS

Erebus ephesperis

Exquisitely patterned

SCIENTIFIC NAME	*Erebus ephesperis* (Hübner, 1827)
FAMILY	Erebidae
NOTABLE FEATURES	Asymmetrical wing pattern
WINGSPAN	3¾ in (96 mm)
SIMILAR SPECIES	Other oriental *Erebus* spp., such as *E. caprimulgus*

Found from China, Japan, and Korea, to Timor and New Guinea, *Erebus ephesperis* is exquisitely patterned with large eyespots, but in shadowy shades of light to dark brown that render it inconspicuous along the rainforest edges and trails of its lowland habitat. The male and female are very similar. Like many of their subfamily Erebinae, the adults feed on the juice of ripened fruits in orchards and tropical forests, and on sap from fallen or damaged trees.

SNAKELIKE CATERPILLARS

The cryptic, orangey-brown *E. ephesperis* caterpillar, which has distinctive eyespot patterning, feeds on the leaves of greenbrier (*Smilax*) vines, as do several other *Erebus* species. When disturbed, it curls its front segments inward to display two prominent, black round patches—a defensive posture, possibly simulating a snake's head and designed to startle a predator. The caterpillar pupates inside a very loose cocoon that incorporates adjacent leaf debris and ecloses two to three weeks later.

ONE OF MANY

The genus *Erebus*, with more than 30 species, includes some impressive moths, such as *E. macrops* of subtropical Africa and Asia with a wingspan of up to 4½ in (120 mm), while in Central and South America, the subfamily Erebinae, which numbers some 10,000 species, harbors real giants of the moth world, such as the Agrippina or White Witch moth (*Thysania agrippina*) with perhaps the world's largest wingspan of almost 12 in (300 mm).

→ A beautiful *Erebus ephesperis* moth rests on the ground, perfectly camouflaged against dead leaves and the earth in Yunnan, China.

Snake mimic

When threatened, an *Erebus ephesperis* caterpillar adopts its defensive, snake-mimicking posture to ward off predators.

Crawling

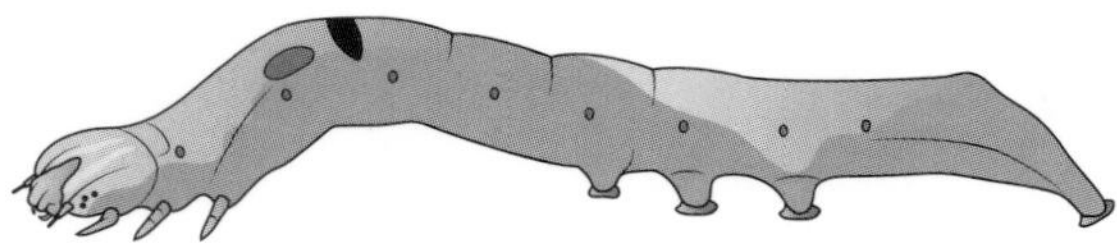

Defensive—simulating a snake head

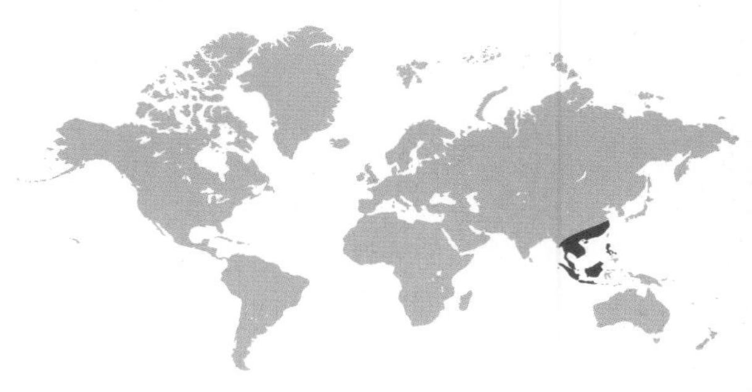

TRABALA PALLIDA

Trabala pallida

Flamboyant yellows and greens

SCIENTIFIC NAME	*Trabala pallida* (Walker, 1855)
FAMILY	Lasiocampidae
NOTABLE FEATURES	Leaf-shaped, bright green males, yellow females
WINGSPAN	1⅓ – 2⅘ in (35–72 mm)
SIMILAR SPECIES	Other *Trabala* spp.

One of 18 species within the genus *Trabala, T. pallida* is a beautiful large moth found in southern Asia from India east to Vietnam and southeast China. Adult males are green, while females are bigger, yellow, and more flamboyant. As its name suggests, the subspecies *Trabala pallida montana*, is found at higher altitudes. At least five *Trabala* species also occur in Equatorial Africa. They are often difficult to identify based on wing pattern alone but, since the mid-twentieth century, as researchers have increasingly studied their internal anatomy, the number of known *Trabala* species has greatly expanded.

Defensive hairs

The head and long, bristly, thoracic hairs of a mature *Trabala pallida* caterpillar.

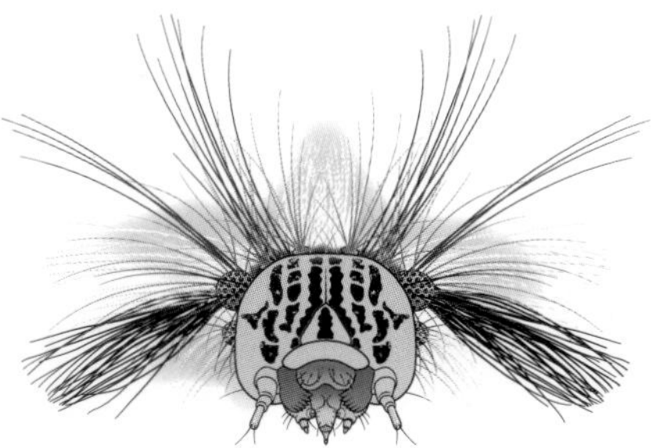

PROTECTIVE BRISTLES

Trabala females lay eggs in clusters, shaped differently according to species, and cover them with hairs from the tufts on their abdomen—sometimes in such a way that clusters can resemble a hirsute caterpillar. The yellowy-brown *T. pallida* larvae have a distinctive white or yellow dorsal line marked with pairs of eyelike tubercles, long hairs branching out behind their heads, and tufts of hairs sprouting sideways from their body. The long, bristly hairs are defensive, as they can irritate the throat of any vertebrate predator that seizes them, prompting their release.

POLYPHAGOUS LARVAE

Trabala caterpillars, like many lasiocampids, are highly polyphagous, feeding on plants including *Lagerstroemia floribunda*, *Melastoma*, *Punica granatum*, *Psidium guayava*, and *Terminalia catappa*. Younger larvae feed in loose groups, frequently defoliating their host tree, and disperse only when close to maturity. When mature, they pupate in a dense, brown, two-humped cocoon, which they attach with silk to a branch of their host plant.

→ A green, leaflike *Trabala pallida* male rests on a branch in Yunnan, China.

BARSINE ORIENTALIS

Oriental Barsine Lichen Moth

Exotic Asian "footman"

SCIENTIFIC NAME	*Barsine orientalis* (Daniel, 1951)
FAMILY	Erebidae
NOTABLE FEATURES	Distinctive red-and-black forewing markings
WINGSPAN	1¼ – 1¾ in (32–45 mm)
SIMILAR SPECIES	Other members of the genus *Barsine*

The exotic *Barsine orientalis* moth flies in Southeast Asia and is one of 65 *Barsine* species, most of which have orangey-red and black speckled and striped forewings. They belong to the tribe Lithosiini, which includes more than 2,700 species, collectively known as "footmen" for their stance when resting, with their front end raised and narrow wings aligned, as if standing to attention. Their similar patterning makes *Barsine* species difficult to identify. Cambodia alone has 12 known *Barsine* species, and new species were described from India as recently as 2020.

A DIET OF LICHENS

Lithosiini are called "lichen moths" because their caterpillars thrive on a diet of lichens—symbiotic (interacting) organisms comprised of fungi, algae, and cyanobacteria, which can grow almost anywhere. It is a tough diet, but ingesting lichen defensive compounds can provide chemical protection for these moths, hence their frequent bright aposematic coloration. Lichen's unique chemicals also enable these moths to develop unique pheromones.

OTHER NOTABLE LICHEN MOTHS

In North America, the Painted Lichen Moth (*Hypoprepia fucosa*), and in Great Britain, the Dew Moth (*Setina irrorella*), are among better known Lithosiini, but there are numerous species elsewhere, especially in the tropics. They frequently mimic other species, sometimes outside Lepidoptera: *H. lampyroides*, described in Arizona in 2018, imitates a toxic firefly. Caterpillars in the lichen moth genus *Cyana*, found from Africa to Australia, use their long hairs to construct spectacular cage-like cocoons, from which the caterpillar manages to eject its skin while pupating.

→ *Barsine orientalis* exhibiting a typical coloration for this large tropical genus, found throughout Southeast Asia.

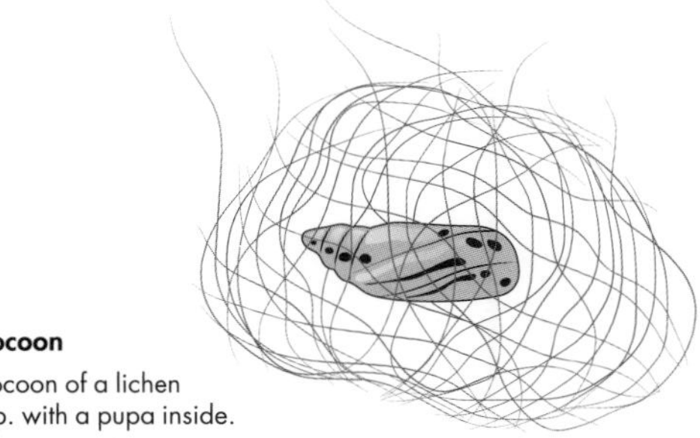

Lichen moth cocoon
The cage-like cocoon of a lichen moth *Cyana* spp. with a pupa inside.

TITULCIA METERYTHRA

Titulcia meterythra

Shimmering wings

SCIENTIFIC NAME	*Titulcia meterythra* (Hampson, 1905)
FAMILY	Nolidae
NOTABLE FEATURES	Yellow forewings, patterned with brown and smaller silver patches, each form a rough equilateral triangle shape when at rest
WINGSPAN	¾ in (19 mm)
SIMILAR SPECIES	Other *Titulcia* spp.; *Ariolica spp.* (though background color and silver patterning differs)

A pretty little moth, with shimmering, silvery patches on its red and yellow wings, *Titulcia meterythra* is one of six *Titulcia* species, all of which occur in Southeast Asia and were first described at the turn of the twentieth century. *Titulcia meterythra* is known from southern China and Malaysia to Borneo and Sumatra.

SIGNALING STRATEGIES

The adults do not feed as they lack a functional proboscis. They have tymbal sound-producing organs in their abdomen, which they may use to signal during courtship or as a defense. Males have hair-pencils (pheromone-signaling structures) within their abdomen, which they extrude to release pheromones during courtship. The shimmering effect of *T. meterythra*'s wings may occur because the gaps between the ridges of each scale are filled and smoothed, making them reflective, as is the case for butterflies, according to recent scanning electron microscopy studies.

BIZARRE LARVAE

Nolid larvae are remarkably diverse—some of them spiny, as in *Nola* species, or similar to cutworms, such as those of the genus *Pseudoips*. Shiny *Titulcia* caterpillars, like those of other members of the subfamily Chloephorinae, have such a swollen, sphere-shaped thorax that a group of larvae resembles a cluster of berries, possibly offering protection by simulating toxic fruits that birds avoid. When disturbed, the caterpillars may drop from leaves on silk threads or flatten themselves and vomit to deter a predator. The larvae pupate openly on the upper surface of leaves in silvery cocoons shaped like upside-down boats.

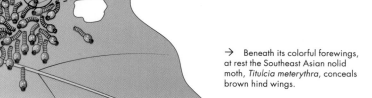

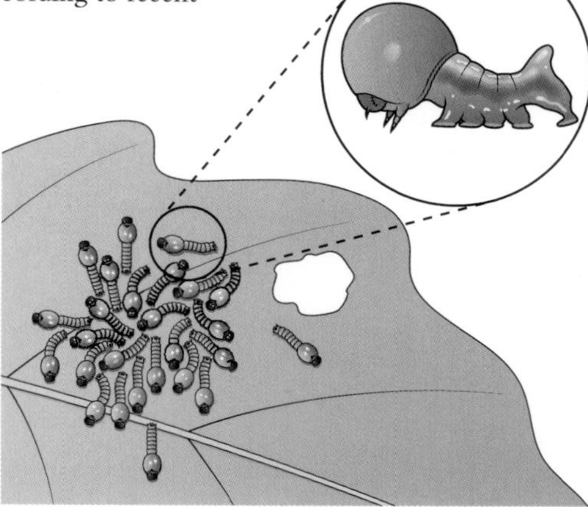

Berry mimic

Larvae of Nolidae such as *Titulcia meterythra* have a uniquely inflated thoracic region that is thought to mimic poisonous berries. When disturbed, they regurgitate liquid (probably noxious), then drop from a plant using a silk line.

→ Beneath its colorful forewings, at rest the Southeast Asian nolid moth, *Titulcia meterythra*, conceals brown hind wings.

URANIA FULGENS

Green Page Moth

Beautifully cryptic nomads

SCIENTIFIC NAME	*Urania fulgens* (Walker, 1854)
FAMILY	Uraniidae
NOTABLE FEATURES	Asymmetrical wing pattern
WINGSPAN	2¾ – 3⅓ in (70–85 mm)
SIMILAR SPECIES	Most similar *Urania* species is *U. leilus*

The day-flying Green Page Moth, often mistaken for a swallowtail butterfly, is one of some 700 mostly tropical moths in the Uraniidae family, which includes both nocturnal and brightly colored diurnal species. While its range extends from Bolivia to the southern United States, *U. fulgens* reproduces only where its *Omphalea* host plants grow in rainforest areas from Veracruz, Mexico, to northern areas of South America.

TOXIC HOSTS

The female lays her eggs exclusively on *Omphalea* spp., which contain toxic alkaloids that confer chemical defenses to the larvae feeding on them and to all stages of the moth. When mature, caterpillars fasten two leaves around themselves and pupate within this shelter.

The adults are also protected by their cryptic coloring—predominantly black with iridescent green lines and stripes on the forewings and smaller green dashes on white-tipped hind wings—which helps to camouflage them when at rest and also signals their toxicity to avian predators. Feeding on nectar from flowers and, in large groups, collecting salt and other nutrients from the ground, males live for an estimated 28 days and females for around 34 days.

NOMADS IN SEARCH OF FOOD

Every four to eight years, entire *U. fulgens* populations—sometimes hundreds of thousands of moths—fly long distances in search of fresh host plants, reaching speeds of up to 13½ mph (21.6 km/h), and crossing expanses of water, such as the Gulf of Mexico. Migrations may occur when their larvae have exhausted the supply of *Omphalea* in one area or because, in response to major *U. fulgens* attacks, *Omphalea* plants increase the toxicity of their chemicals to levels that the larvae cannot tolerate.

→ The Green Page Moth in Costa Rica; this species displays iridescent green markings created by light reflected at a particular wavelength.

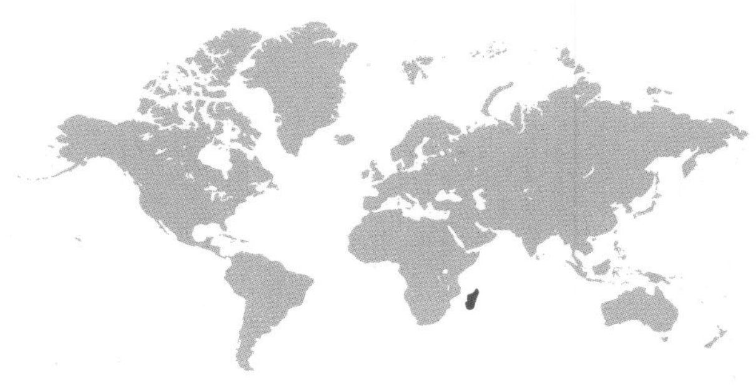

ARGEMA MITTREI

Comet Moth

Long tails and dramatic eyespots

SCIENTIFIC NAME	*Argema mittrei* (Guérin-Méneville, 1847)
FAMILY	Saturniidae
NOTABLE FEATURES	Long tails, especially in males
WINGSPAN	3⅛ – 4¾ in (80–120 mm)
SIMILAR SPECIES	Smaller African Moon Moth *(Argema mimosae)*, found from East to South Africa

Native to the residual rainforests of Madagascar, the Comet Moth is striking for the elegant beauty of its extensive tailed hind wings. Both sexes are a vibrant yellow, patterned with red frills and dramatic red eyespots, reflected in the genus name *Argema*—Greek for "speckled eye." Their patterning and shape are part of their defenses if attacked. A flash of the eyespots can startle a predator, while the spinning tails confuse bats at night, which might aim for these, rather than more vulnerable parts of the moth's anatomy.

SEXUAL DIMORPHISM

The male is around 8 in (200 mm) in length including the "tails"—the tapering red hind wing extensions with widened yellow tips. Its large, feathery antennae enable him to detect the female's pheromones. As in many moths, the females, which have shorter hind wings, are larger, heavier, and more sedentary. When they emerge from pupae, they are already full of unfertilized but fully formed eggs and can each lay more than 150. Like all other saturniids, these moths do not feed as adults and live less than a week.

BRED FOR TRADE

The Comet Moth's bright green caterpillars develop through four instars feeding on *Weinmannia eriocampa* and *Uapaca*, native to Madagascar, and on introduced eucalyptus. They then spin silken cocoons, peppered with tiny holes so water can drain through, ensuring survival in heavy rainfall during the five- to six-month pupation period. The Comet Moth is among the handful of moth species that are bred for trade as a sustainable way of providing income for the local people in Madagascar. While promoting insect conservation on the island, the cocoons are shipped around the world, and the resulting moths are displayed in insect exhibits.

→ The Comet Moth, endemic to Madagascar, with its long-tailed hind wings that help deflect bat attacks.

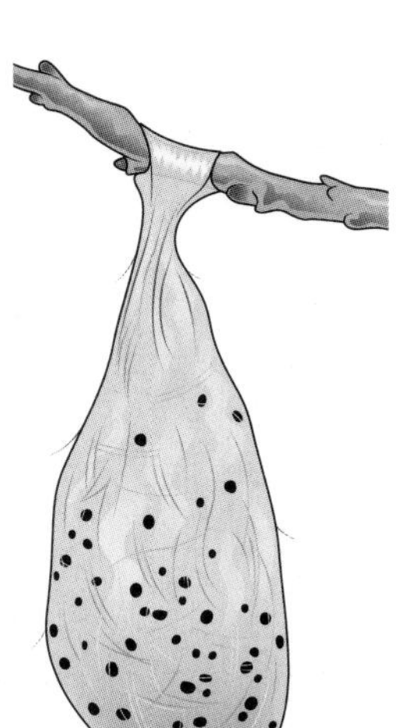

Air pockets

The tiny holes in the Comet Moth's cocoon enable water to drain through, preventing the pupa from drowning during heavy rains. The shiny cocoon surface reflects much of incoming solar radiation—a cooling method so highly efficient that the use of something similar has been explored by researchers in cooling clothing technology.

ROTHSCHILDIA ERYCINA

Rothschildia erycina

Colorful and intricately patterned

SCIENTIFIC NAME	*Rothschildia erycina* (Rothschild, 1907)
FAMILY	Saturniidae
NOTABLE FEATURES	Large, clear membrane "windows" in wings, extended wing tips in males
WINGSPAN	3½ – 4¾ in (90–120 mm)
SIMILAR SPECIES	28 other *Rothschildia* spp.

Among the most colorful and intricately patterned silk moths, *Rothschildia erycina* (divided into several subspecies) is widespread from southern Mexico to Brazil and produces several generations per year. These moths do not feed as adults and have a lifespan of only 7 to 14 days. The females, which can lay more than 100 eggs, are considerably larger than the males, and the tips of their forewings are more rounded.

APOSEMATIC CATERPILLARS

Banded yellow and green when young, the larvae live initially in groups, but become solitary at later instars when their variable coloring includes thick bands of black, adorned with bright orange blotches and small tufts of hair. The caterpillars are polyphagous, but feed conspicuously on toxic plants such as *Ailanthus*, *Coutarea*, *Exostema*, and *Antonia* in the Rubiaceae and Loganiaceae families, from which they may acquire a subtle chemical protection. While there is no direct evidence of their toxicity, if disturbed, the larvae can vomit a substance that may be irritating to predators. When mature, they spin oval cocoons, suspending them from branches with their silk.

ROTHSCHILDIA SILKS

Historically, Native Americans in Arizona used *Rothschildia* cocoons to make culturally significant objects, such as rattles and musical instruments. Since pre-Columbian days, people in remote northwestern areas of Argentina have collected *Rothschildia* cocoons for their silk and are now encouraged to take them after the moths have eclosed, creating a sustainable local industry.

→ A male of *Rothschildia erycina*.

Female wing tips—more rounded

Male wing tips—more extended

Spotting the difference

Though similar in their wing pattern coloration, males and females of *Rothschildia erycina* have different wing shapes.

MOTHS OF GRASSLANDS & MEADOWS

Diverse grassland ecosystems

Covering as much as 40 percent of Earth's habitable land, grasslands are one of the world's largest ecosystems. They include the great Brazilian Cerrado, North American prairies, the steppes of Eurasia, the tropical savannas of sub-Saharan Africa and northern Australia, and also areas between the tree line and snow line on most high mountains around the world. They may appear to lack the variety of plant and moth species found in rainforests, but conservation biologists have recently shown that some areas of grassland can be exceptionally rich and diverse habitats, harboring moth species found nowhere else.

EXPANDING GRASSLANDS

Grasslands have much greater daily and seasonal temperature fluctuations than adjacent forests. Like forests, they have dominant plant species, but here these are grasses and a wide variety of herbs and shrubs that serve as host plants and form unique associations with moth fauna. For the past 5,000 to 7,000 years,

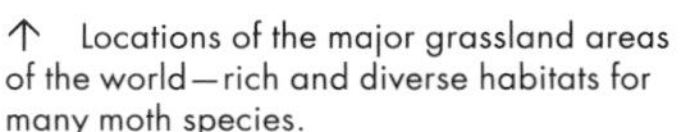

↑ Locations of the major grassland areas of the world—rich and diverse habitats for many moth species.

← Alpine meadow habitat in the Rocky Mountains, United States; mountain grasslands are home to some unique moth species.

as humans have cut down trees to gain land to live and grow food on, grasslands have expanded—a process that has greatly accelerated in recent centuries with exponential growth in human populations. Intensive, unsustainable agriculture now poses a severe threat—not only to the world's forests, but also to some of the grassland biomes and their indigenous wildlife species.

TROPICAL SAVANNAS

Sandwiched between deserts and rainforests, tropical savannas enjoy year-round warmth, with temperatures of 68–78°F (20–25°C) in winter and 78–86°F (25–30°C) in summer, but have a sharply contrasting rainy, humid wet season and harsh, dry season when fires frequently occur. Plants have adapted to these extremes; grasses are often dormant in the dry season, and trees shed their leaves. Some trees have developed a thick bark to help withstand fires, while others, such as baobabs, absorb water during the wet season and store it in their trunks. The moths that depend on these plants have similarly adapted with a variety of survival strategies.

↖ A freshly emerged Nine-spotted Moth (*Amata phegea*)—a widespread European moth, frequently found in meadows.

↗ A female Ghost Moth (*Hepialus humuli*) whose larvae feed on roots of a variety of herbaceous plants. Its name derives from the ghostlike, hovering flight of adult males during mating displays.

Moths of the Cerrado

The Brazilian Cerrado, the largest savanna region of South America, which is home to rare wildlife such as the maned wolf and jaguar, includes at least 1,000 tree and shrub species, and more than a third of its plants grow nowhere else in the world.

ABUNDANT SPECIES

The numerous grass species (Poaceae) include beard grasses (*Andropogon*) and carpet grasses (*Axonopus*), which differ between the dry Cerrado areas and wetter parts. Dry areas are also often characterized by fan grasses (*Eustachys*), skeletongrass (*Gymnopogon*), beautiful Indian grass (*Sorghastrum*), and elegant *Loudetiopsis* grass, while tall *Panicum* grass, up to 10 ft (3 m) high, and Bahia grasses (*Paspalum*) are among the dominant species in the wetter parts. The range of moths is correspondingly large. For example, surveys here have revealed 723 species of tiger moth (Arctiinae, subfamily of Erebidae), with the highest local occurrence of around 200 species while, by comparison, the tiger moth diversity of the entire United States barely exceeds 300 species.

PROLIFIC POLLINATORS

Moths play an important ecological role in the Cerrado. A detailed 2004 study revealed that they pollinate a number of ecologically and economically important plant species, which often have flower structures exclusively adapted to moth pollination. Among such plants are *Alibertia edulis*, whose fruit is used locally to make jam, *Roupala montana*, which is important for firewood and construction, and two species of timber trees, *Aspidosperma* and *Diospyros*.

The vivid red flowers of *Ferdinandusa* trees are pollinated by hawk moths, as are those of the mangabeira tree (*Hancornia speciosa*), which produces sweet, yellow-red mangaba fruits, used for flavoring juices, ice cream, preserves, and wine. The flowers of these trees and of many other plants pollinated by hawk moths, have narrow openings and long corollas—the floral tubes that only the proboscis of a hawk moth can enter. Many of the plants that hawk moths pollinate, such as decorative *Qualea* and *Salvertia* spp. (Vochysiaceae) are found exclusively

in dry areas of savannas and are ecologically important because they are among the first to re-emerge when habitat is destroyed by fire.

↑ Recently burned woody vegetation growing back in the Brazilian Cerrado on the hills of Capitólio, Minas Gerais state.

→ The Cerrado is home to numerous hawk moth species, among which is the Pink-spotted Hawk Moth (Agrius cingulata), a widespread, mostly Neotropical species that, as a caterpillar, feeds on morning glories and Datura spp., and pollinates numerous flowers with deep corollas.

It is not only hawk moths that have an intimate connection with the Cerrado's unique flora. Among the 400 species of silk moths (Saturniidae) that fly here (twice as many as fly in the United States), 160 feed only on this habitat's vegetation. Given the diversity in these two families, researchers estimate that the Cerrado's moth species may number as many as 20,000. Encroaching agriculture, however, is increasingly affecting their habitat: by some estimates only about 20 percent of the land is now undeveloped, and 2020 fires were at least as devastating here as they were in the Amazonian rainforest.

CONSERVING BIODIVERSITY

After a long decline in natural history expeditions from the middle of the twentieth century, Brazil now has numerous research groups of talented, well-trained researchers, who conduct detailed local studies demonstrating the uniqueness and importance of moth fauna for biomes such as the Cerrado. Thanks to the Internet, sharing information has become much easier, although research and public policies are not as strongly connected as they must be to conserve biodiversity for future generations. However, awareness is the first step of conservation, and the existence of good basic research in Brazil raises hope that some of the region's biodiversity will survive the current destruction wrought by agriculture and development.

↑ Many rainforest Saturniidae species, such as *Rhescyntis hippodamia*, can be found in the Cerrado thanks to woody tropical vegetation that grows along rivers which run through the biome.

← The Tetrio Sphinx (*Pseudosphinx tetrio*) may have dull-colored adults, but its caterpillars are very colorful. In the Brazilian Cerrado, the moth has been observed to pollinate the "souari nut" tree (*Caryocar nuciferum*), which has an edible fruit popular in Central and Western Brazil.

DISCOVERIES IN THE CERRADO

In 1758, in the 10th edition of his *Systema Naturae*, the Swedish taxonomist, botanist, zoologist, and physician Carl Linnaeus formally described two moth species that occur in the Cerrado among other places—the Ornate Bella Moth (*Utetheisa ornatrix*), right, and the wasp-mimicking *Saurita cassandra*. During the nineteenth and early twentieth centuries, entomologists such as William Schaus of New York and, in London, Francis Walker, George Francis Hampson, and Walter, the colorful second Baron Rothschild, were responsible for most of the descriptions of Cerrado moths. Their work was based on specimens shipped from South America by dedicated collectors, often sponsored by societies or figures like Rothschild. Many type specimens were later housed in major European museums. For example, Hampson's catalog of moth type specimens stored in the Natural History Museum in London is still one of the main identification guides used, more than 100 years after its publication. The choice of moths collected was often not random; larger and more showy species were described first, which is one reason why more is known about some groups of moths rather than others.

The African savanna

Savanna covers 5 million sq miles (13 million sq km) of Africa. Its rich grassland and drought-tolerant trees, such as African teaks and acacias, support many large herbivores, including zebras, elephants, giraffes, and a surprising diversity of moths.

OUT OF AFRICA

Kenya's savanna, which has two rainy seasons annually—in spring and in fall, with a total rainfall of up to 50 in (1,270 mm)—is home to big cats, cheetahs, and lions, and also spectacular moth species, such as the large Cream-striped Owl Moth (*Cyligramma latona*) with its striking forewing eyespots and the Oleander Hawk Moth (*Daphnis nerii*), whose camouflaged, marbled-green pattern resembles polished malachite. Cream-striped Owl Moth caterpillars feed on acacia trees, while Oleander Hawk Moth larvae feed on oleander, the now widespread ornamental plant that gives the species its name. Both moths migrate long distances in search of their host plants, and the Oleander Hawk Moth flies as far as Ukraine, north India, and China.

ANCIENT ASSOCIATIONS

Certain ancient moth associations may have developed on Africa's savanna, such as those between rattlebox (*Crotalaria*) plant species and tiger moths (Arctiinae), whose larvae feed on these plants and ingest their toxic alkaloids, which protect all moth stages against predators. Africa has more than 400 species of *Crotalaria*, which evolved on that continent 23 to 30 million years ago. They now grow mainly on damp grasslands and are often used as cover crops (non-cash crops that protect or improve land between regular crop production). Researchers believe that tiger moths, which feed on *Crotalaria*, may have originated here and then become widespread and greatly diversified around the world. In Africa, these species are represented by the Cheetah (*Argina amanda*) and the Beautiful Tiger (*Amphicallia bellatrix*) among others. The Crimson-speckled Flunkey (*Utetheisa pulchella*), whose close New World relative, the Ornate Bella Moth (*U. ornatrix*) also feeds on rattlebox plants, feeds on other herbaceous alkaloid-containing plants throughout Africa.

FUNGAL RELATIONSHIPS

A number of moth species have a fascinating association with a fungus on the beautiful sweet thorn tree (*Vachellia karroo*), which grows in South Africa from the Western Cape to Zambia and Angola, and is an indicator of temperate grassland known as sweet veld, which is especially good for grazing. Moths and butterflies are among the insects that pollinate the tree's yellow pompom flowers, and the caterpillars of more than 20 moth species tunnel into galls (swellings or abnormal growths) created on the tree by the rust fungus *Ravenelia macowaniana*, which offer nutrients, shelter, and protection from parasitoids and predators. Similar associations exist in Australia, involving galls of the fungus *Uromycladium tepperianum* on the widespread black wattle trees (*Acacia decurrens*), which shelter larvae of seven species of moths in the families Tortricidae, Tineidae, Gracillariidae, Pyralidae, and Stathmopodidae.

↑ Thelwall's Beautiful Tiger (*Amphicallia thelwalli*), an arctiine moth, here featured on a Mozambique postage stamp.

← The Crimson-speckled Flunkey (*Utetheisa pulchella*) is common throughout the grasslands of Africa and Europe.

←← African savanna at sunset—an open landscape dominated by grass and scattered with trees, where moth caterpillars play their part in a food web as herbivores and food for birds.

SEASONAL ADAPTATIONS

Rainfall in the southern African savanna is lower than in Kenya. Grasslands bordering the Kalahari Desert may have only 4 in (100 mm) of rain during the wet season. Moth life cycles here, as elsewhere, are rain-dependent, as water gives rise to new vegetation for caterpillars to feed on, and causes a brief bloom of flowers that are visited by adult moths. Despite the arid conditions, certain moth species appear to flourish. One brief survey of several South African grassland parks revealed 70 species of plume moths (Pterophoridae)—around half of the known Pterophoridae species in the United States and Canada combined. The tiny adults have featherlike wings and a pale brown coloring that blends in with the dry, grassy vegetation, and their larvae feed on drought-tolerant plants. Researchers suspect that further studies and surveys will reveal that other micromoths are equally abundant in this habitat.

CATERPILLAR DELICACIES

Numerous caterpillars appear on the savanna in spring. Most are common species, such as the large Mopane Worm (*Gonimbrasia bellina*), which feeds primarily (but not exclusively) on the mopane tree (*Colophospermum mopane*) that grows in savanna woodland across southern Africa—even in areas with 4 in (100 mm) rainfall annually. A good number of mopane larvae never reach adulthood as they are an important source of protein for both wildlife and humans. A popular delicacy, they are handpicked in the wild, boiled and salted, then dried in the sun or smoked, and sometimes industrially canned.

Among the larger and showier, but very common moths of African savanna is the Pallid Emperor Moth (*Cirina forda*), whose caterpillars feed on the wild syringa (*Burkea africana*) and the shea tree (*Vitellaria paradoxa*). This beautiful silk moth pupates in soil

← A late-instar caterpillar of the Pallid Emperor Moth (*Cirina forda*). The larvae can defoliate shea trees in West Africa.

↙ Lekkerbreek tree or peeling plane tree (*Ochna pulchra*)—host plant to *Bostra pyroxantha* pyralid moth larvae in South African grassland.

↓ Roasted Mopane Worms—caterpillars of the saturniid *Gonimbrasia bellina*—are a common food staple in South Africa.

underneath the plant, and its large caterpillars are, like Mopane Worms, a source of nutrition for both wildlife and local people. Dried caterpillars are a staple in southwestern Nigeria and especially important in children's diets as they contain 50 percent protein and 17 percent fat as well as calcium, iron, and zinc.

Further common and ecologically important savanna species include the Banded Euproctis (*Knappetra fasciata*), a tussock moth whose caterpillars are protected by urticant (stinging) hairs and feed in groups on many species of trees, frequently defoliating them. Birds and other wildlife prefer more palatable noctuid caterpillars, such as Arch Drab (*Maurilia arcuata*) and *Neaxestis piperitella*, which feed on clusterleaf (*Terminalia sericea*). Together the caterpillars of these two species transform as much as 4 percent of the dry leaf biomass into fats and proteins, which are passed onto the birds that feed on them, or turned into frass (dung) that is also important for plants and animals in the ecosystem. Caterpillars of the brightly colored pyralid moth, *Bostra pyroxantha*, defoliate the lovely peeling plane tree (*Ochna pulchra*), also known as lekkerbreek (which in Afrikaans means "break easy") because of its brittle branches.

The Australian savanna

The tropical savanna of northern Australia, an area of more than 500,000 sq miles (1.3 million sq km) covering the northern section of Western Australia, the Northern Territory, and Queensland, is grassland dotted with scattered eucalyptus of different species with exotic names such as bimble box, cuoolibah, red river gum, and black box. While heavy grazing by livestock has greatly changed the nature of this habitat, its remaining moth populations are associated with eucalyptus, kangaroo grass, and other plants that still grow here.

EVOLVING THROUGH ISOLATION

The wildlife of Australia is largely endemic due to its prolonged isolation from the rest of the world and much of its moth population is dependent on any one or several of some 700 eucalyptus species. In one brief moth survey covering 300 species, 70 percent fed on eucalyptus. For some families, such as Oecophoridae, the percentage is even higher, and their caterpillars feed not only on fresh leaves, but also on fallen dry leaves. Notable moths of Australian grasslands include the Golden Sun Moth (*Synemon plana*), now critically endangered, and one of 24 day-flying Australian castniids. The larvae of these moths feed on the roots of wallaby grass (*Austrodanthonia*). Golden Sun Moth females lay around 200 eggs at the base of their host plant. After three weeks, the eggs hatch, and the larvae tunnel through grass roots feeding on them as they develop. Adult females emerge with fully developed eggs and attract their diurnal mates with their bright orange wings. The adults do not feed and are short-lived, producing only one generation per year in this harsh habitat.

GHOST GIANTS AND OTHER ENDEMICS

Tunneling underground, including through the roots of grasses, is common in the primitive moth family of Ghost Moths (Hepialidae), which is well represented in Australia. In 2017, a study revealed 15 new species and

one new genus of Australian ghost moths. Some of Australia's largest and most spectacular moths belong to this family, with species reaching 6¼ in (160 mm) in wingspan. Other species found in Australian grasslands include large, beautiful Australian lappet moths, recently reclassified into a separate family—Anthelidae. The continent has more than 70 Anthelidae species, including the Variable Anthelid (*Anthela varia*), from coastal areas in the south, whose caterpillar, nicknamed "hairy mary," feeds on bushnut (macadamia), eucalyptus, toothbrush plant (*Grevillea*), and the firewheel tree (*Stenocarpus sinuatus*).

↖ *Abantiades labyrinthicus* (Hepialidae) is a ghost moth endemic to Australia; it reaches a wingspan of 6⅓ in (160 mm).

↑ The Golden Sun Moth (*Synemon plana*), found in grasslands, is one of several of the Australian castniids. A male (here) flies by day looking for a sedentary female, which is less mobile and larger, with relatively smaller hind wings.

← Grassland habitat in Windjana Gorge, Kimberley region of Western Australia, favored by sun moth (Castniidae) species.

Grassland "islands" in the mountains and meadows

Mountain grasslands occur around the world above the tree line and below the snow line, and are created by the harsher climate and poorer soil at high altitudes. While such grasslands are frequently dry, sometimes they turn into lush alpine meadows, fed by streams formed by melting snowcaps and glaciers. Because these grasslands are all quite separate, they are thought to include a number of unique moth communities but are only just beginning to be explored.

MOTHS OF THE HIGH ANDES

Above the tree line of the Andes of Colombia and Ecuador, at elevations of over 9,000 ft (3000 m), lies the vast Northern Andean páramo. From 10,000 years ago, as the climate warmed, the páramo and the trees below it migrated steadily upward and separated into smaller isolated grassland areas that contain rare and endangered wildlife and a vast number of insects, including moths. It is cold on the mountain grasslands at night, so many moths become diurnal, and various micromoths, such as tortricids, yponomeutoids, and pyraloids, fly just after sunset before it gets too cold, although some species, especially noctuids and geometrids, still fly at night even when the temperature is below 46°F (8°C).

In this unusual habitat new species and subspecies have evolved. However, unlike Andean tropical forests below, whose diversity may be concentrated in numerous ecological niches within a single square mile, the diversity of the páramo is spread over 4,350 miles (7,000 km) of the Andean mountain chain, where rare highland plant and corresponding moth communities are isolated. A recent survey of leaf miners in the family Nepticulidae revealed a vast undescribed diversity.

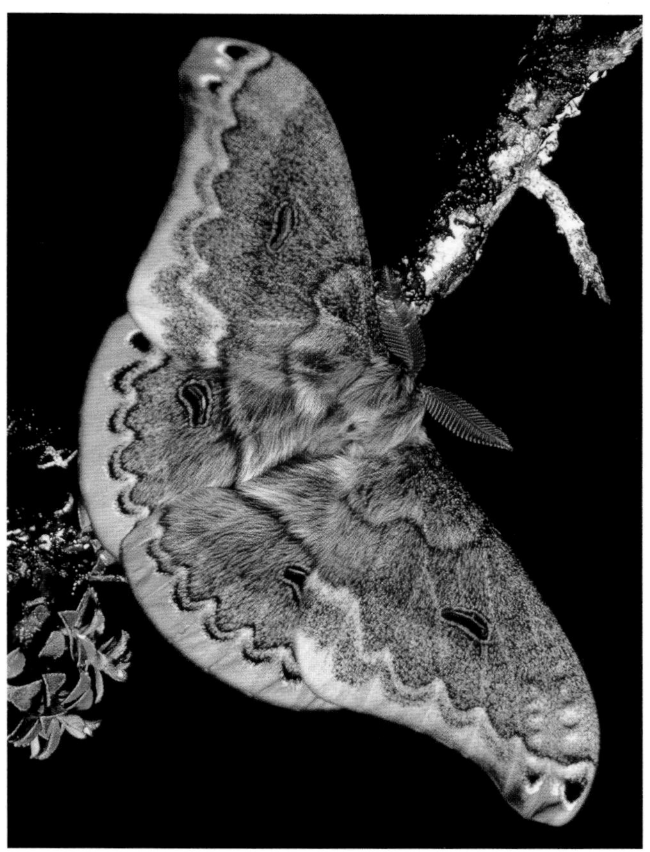

↗ A *Copaxa orientalis* silk moth photographed around the Tungurahua volcano, more than 13,100 ft (4000 m) above sea level, in the Cordillera Oriental of Ecuador.

← Wild vicuñas—camel relatives—graze on sparse grassland in the highlands of Ecuador around Chimborazo Mountain, in the Cordillera Occidental of the Andes, where endemic moth communities have developed close to their rare, isolated host plants.

MOTHS THAT LIVE IN TORTOISE BURROWS

In grassy areas along forest edges or open, sandy, upland pine habitats of the southeastern United States, gopher tortoises create long, deep burrows, which are home to other creatures, too, including three families of moths. Larvae of the tineid moth *Ceratophaga vicinella* feed on the shells of dead gopher tortoises, constructing silk tubes that anchor the shell to the soil, and later pupate within these tubes. In a 2019 study in Florida scrub habitat, little snow-white *Acrolophus pholeter* moths were seen emerging from three out of the five gopher tortoise burrows examined. *Acrolophus pholeter* larvae can take over a year to develop on gopher tortoise droppings and decaying plant matter. Caterpillars of the noctuid Tortoise Commensal Noctuid Moth (*Idia gopheri*) also live in gopher tortoise burrows and probably feed on fungus growing on tortoise droppings and other litter.

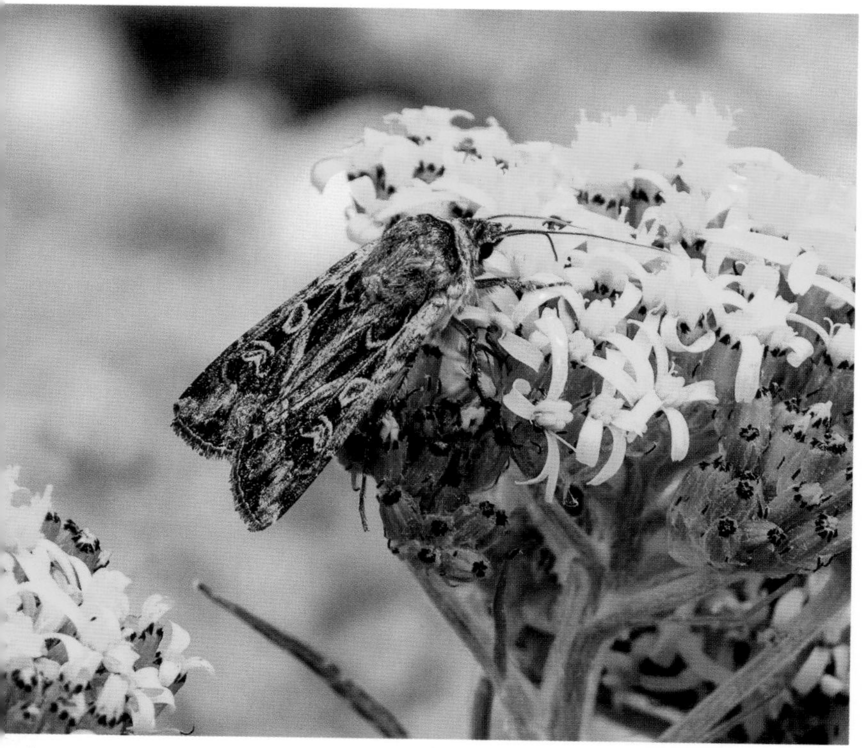

← Army Cutworm (*Euxoa auxiliaris*) moths migrate from the hot plains of western North America into cool mountains, but may be preyed on there by grizzly bears.

→ View of the Australian Alps from the Mount Buffalo National Park in Victoria, where maximum summer temperatures average around 73°F (23°C).

GRASSLAND MIGRANTS

Patterns of diversity similar to those of the South American páramo are found along all high mountain chains of the world, such as the Himalayas and the Pamir in Asia, or the Alps and the Pyrenees of Europe—each with moth communities of their own. But in some of these habitats, unique moth behaviors have been described. One of these phenomena is migration.

After they complete their development in the spring and early summer, Army Cutworm (*Euxoa auxiliaris*) moths migrate from the hot, dry grasslands of North America's Great Plains, where their caterpillars feed on grasses (including agricultural crops) and other herbaceous plants. The moths eventually reach the alpine meadows of the Rocky Mountains, where they sip on flower nectar, and, like migrating monarch butterflies, accumulate fats that will come in useful during hibernation. Once fortified, they continue onto the high plateaus of the Rockies looking for sites where they can estivate (hide from the summer heat in a dormant state) at a cool temperature during hot summers. These habitats include the alpine tundra of Yellowstone National Park in the western United States. There, the moths are preyed on by grizzly bears, who greatly rely on this food source during the summer. The bears follow the moths, migrating upward from the forests, and turn over rocks to find them, consuming them by the tens of thousands. The surviving moths return to the plains when they are ready to reproduce in the fall.

LONG MIGRATIONS

The Australian Bogong Moths (*Agrotis infusa*) are notable for their long migrations—up to 600 miles (965 km)—to escape seasonal heat. Like cutworms, their larvae are widespread in spring and damage crops but, unlike cutworms who favor grasses, they prefer capeweeds, cabbages, peas, and potatoes. As adults, they fly by the millions to the southern Australian Alps of eastern Victoria and southeastern New South Wales, where they, too, estivate at high elevations, hiding from sunlight in large social clusters of up to 1,600 moths per sq ft (17,000 moths per sq m).

EATING MOTHS

Anthropological evidence suggests that Australian Aboriginal peoples have also dined on adult Bogong Moths for at least two millennia, and would travel up to the Australian Alps to take advantage of this resource, celebrating moth estivation with seasonal festivities and inter-tribal gatherings. The moths were cooked into cakes and were said to have a nutty flavor. Eating Bogong Moths, however, now comes with a health warning: as a result of earlier use of pesticides containing arsenic in the grazing and cotton-growing areas where the moths feed, the insects can ingest and retain traces of the dangerous chemical.

↑ Bogong Moths (*Agrotis infusa*) migrate seasonally to the Australian Alps where they estivate (spend the summer) in cooler caves and under rocks, congregating in groups of tens of thousands.

Moths in the meadows

As one walks through a meadow during the day, not only butterflies, but also moths will often fly up. Some that fly in early spring, or at high elevations, or in colder climates are active during the day to avoid low temperatures, but there can be a variety of reasons for the daytime moth activity. Many of these moths have diurnal lifestyles, and, like butterflies, may take advantage of diurnal flowers and be toxic to predators. Others are simply active both by day and by night, or are "light sleepers," always ready to fly if disturbed. Depending on biology, the moths found in meadows may be cryptically colored or colorful.

NOT ALL THAT GLITTERS

Metallic coloration is rare in Lepidoptera and is entirely structural. It occurs at a nano-level as the grooves on the scales are smoothed out so that they become more reflective compared to the more common non-metallic scales. Notable species of open grassy areas include noctuids of the subfamily Plusiinae, such as the Silver Y (*Autographa gamma*) moths, frequently characterized by silver and gold metallic markings on their forewings. These moths migrate to northern Europe and the British Isles from their breeding grounds in southern Europe, sometimes reaching high numbers, and three generations annually arrive in a series of waves. Silver Y moths feed as caterpillars on a wide range of herbaceous plants, such as clover, nettles, peas, and various crucifers, including cabbage. They are an alien, invasive species in the United States, but in their native range they play an important role in their ecosystems, flying swiftly by day and by night and pollinating a variety of flowers. There are at least 40 different *Autographa* species around the world and numerous related plusiine genera with similar biology and appearance. Their caterpillars are referred to as loopers, resembling inchworms with their characteristic earth-measuring gait.

Among plusiines are the Burnished Brass (*Diachrysia chrysitis*) moth and its relatives. They have a spectacular metallic green forewing coloration, sometimes spanning almost the entire wing, which attracts the attention of many moth watchers and photographers. The Burnished Brass moth, which occurs across Europe, also prefers open areas, and feeds as a caterpillar on herbaceous plants such as nettles, thistles, and oregano. In the United Kingdom, its distribution and when it flies (phenology) are affected by the current rapid change in climate, as is the case for many butterfly and moth species.

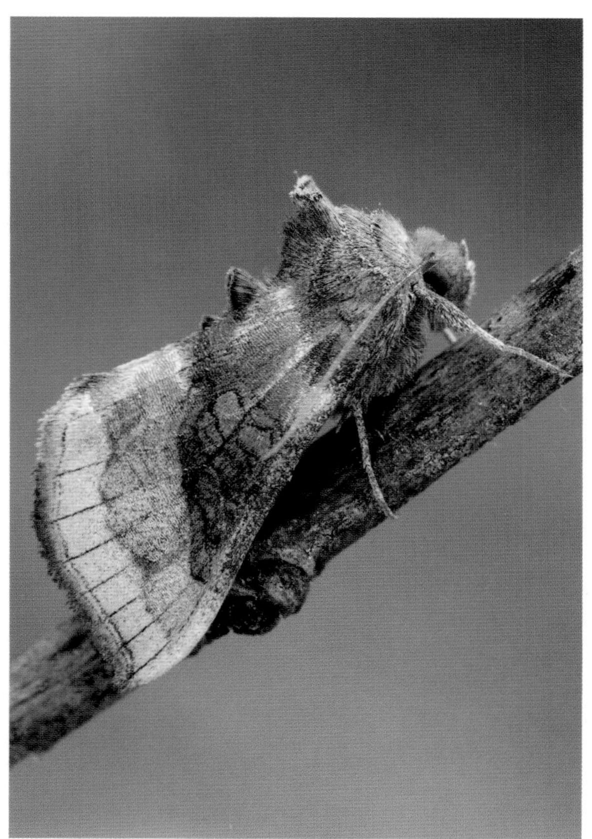

↑ Hummingbird Hawk Moths (*Macroglossum stellatarum*) sip nectar from verbena flowers.

← A Burnished Brass moth (*Diachrysia chrysitis*) rests, well camouflaged by its coloring, on a twig in a German meadow.

←← The Silver Y (*Autographa gamma*) is widespread over almost all of Europe, northern Asia, and North Africa. Its larvae feed on over 200 species of herbaceous plants including those of the pea and cabbage family.

TOXIC BEAUTIES

Some colorful day-flying grassland moths have no need to fly rapidly or camouflage themselves, because their bright reds and yellows indicate that they are well defended from predators by their toxic chemicals. Burnet moth caterpillars of the genus *Zygaena*, for example, sequester cyanogenic compounds from host plants such as the bird's-foot trefoil (*Lotus corniculatus*). These moths are common in grasslands in temperate Eurasia and North Africa, and frequently spotted in the alpine habitats of the Pyrenees or the Alps. Many of them, such as the Mountain Burnet (*Zygaena exulans*), fly only at high elevations and are more common in moorlands, while others, such as the Six-spot Burnet (*Zygaena filipendulae*), span a wider range of elevations, showing remarkable adaptability. All *Zygaena* spp. rest on flowers and are very easy to catch; as a result, these colorful moths have been popular for centuries with butterfly collectors. Moths similar to *Zygaena* fly in Africa but have exotic genera names such as *Zutulba* and *Saliunca*.

Another colorful, toxic group of wasp-mimicking moths of the erebid genus *Amata*, which includes more than 150 species, are found in grassy habitats across the world. The Maid Alice (*Amata alicia*), flies in savanna habitat of sub-Saharan Africa, with larvae (called woolly bears for their hairy appearance) feeding on several plants, such as *Bidens*. The woolly bear caterpillars of other *Amata* species, such as the Nine-spotted Moth (*A. phegea*) in Europe, feed on a variety of herbaceous

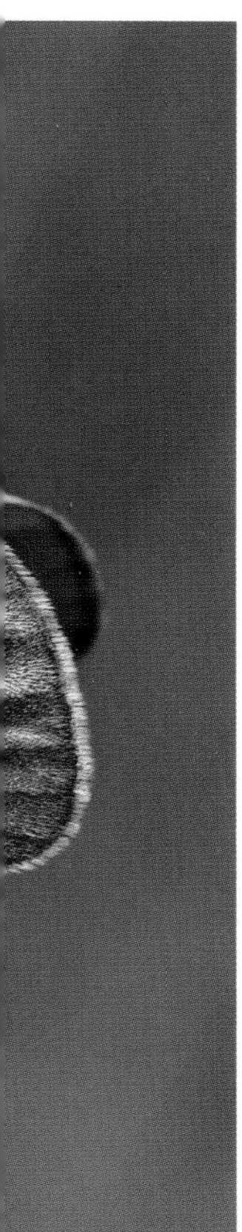

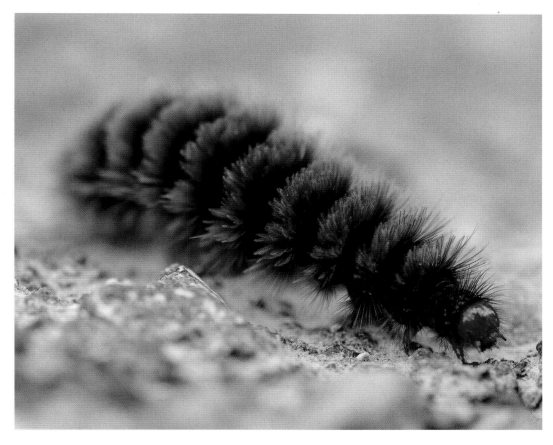

← Six-spot Burnet (*Zygaena filipendulae*) caterpillar feeding on lotus from which these larvae (and the resulting adults) gain chemical protection.

↑ A Mountain Burnet (*Zygaena exulans*) imbibes nectar from a flower in Val de Bagnes, a valley in southwest Switzerland.

↗ A mature woolly bear caterpillar of the Nine-spotted Moth (*Amata phegea*).

plants such as sorrels, dandelions, and grasses, while Hübner's Wasp Moth (*A. huebneri*) larvae in Australia also favor decomposing leaves and Asian rice. Although the host plants of Nine-spotted Moth caterpillars contain no toxic alkaloids, the moths still produce a histamine-like substance that may assist in their chemical defense. Males of these moths, like most of their relatives, have androconial brushes on their front legs that help them disperse pheromones during mating.

CAMOUFLAGE AND MIMICRY

When disturbed by day in meadows, whether resting or nectaring on flowers, delicate light and dark brown geometrid and noctuid moths fly a few yards, settle, and disappear from view. The moment they land in their grassy habitat, they become hidden by the perfect match of their color and pattern to the background of dry grass stems and leaves.

More than 120 species of the *Schinia* moths, mostly found on North American grasslands and meadows, are so well camouflaged that they have no need to fly up and hide. Even though they are frequently active by day, most people are unable to spot them. *Schinia* moths spend much of their time sitting on flowers—hence their nickname of "flower moths"—and their coloration often closely resembles the flowers they visit, with lines in their patterning even simulating the shadows created by the shapes of petals.

HOVERING HAWK MOTHS

By day and by night, hawk moths pollinate meadow flowers. Day-flyers in Europe and North America include bee hawk moths (*Hemaris* spp.), which take nectar from a variety of flowers from lantana and pentas to bugle and valerian. Hummingbird Hawk Moths (*Macroglossum stellatarum*), which visit valerian,

→ The Broad-bordered Bee Hawk Moth (*Hemaris fuciformis*) at rest in a meadow in Switzerland.

↙ Primrose Moths (*Schinia florida*) feeding and resting on an evening primrose flower, which also serves as food for their larvae.

↓ The Coffee Bee Hawk Moth (*Cephonodes hylas*) is widespread in warmer areas of the Old World—here sipping nectar while hovering over a cosmos flower.

honeysuckle, and jasmine flowers among many others, are found in southern parts of Europe, flying south to the African north coasts, India, and Southeast Asia in the winter, and all the way to northern Europe during the summer. A similar species, the African Hummingbird Hawk Moth (*M. trochilus*), is found in southern Africa. Caterpillars of the Hummingbird Hawk Moths feed on herbaceous plants such as bedstraws (*Galium* spp.) and wild madder (*Rubia peregrina*). While these and other small diurnal hawk moth species may be mimicking bumblebees, presumably to avoid being eaten by birds, their flight is more similar to that of hummingbirds. They hover over flowers and are almost never observed landing. This high-energy flying style enables the moths to move swiftly and maintain a relatively constant high body temperature, regardless of the outside air temperature. Like hummingbirds, many of the hawk-moth species seen in grasslands, meadows, and deserts are migratory, and breed in different habitats in different seasons.

A range of temperate grasslands

While there are generalist species in temperate grasslands that arrive from adjacent forests, there are also some specialists found only in these areas; these species feed on unique flora that grows sporadically on special soils within these vast spaces.

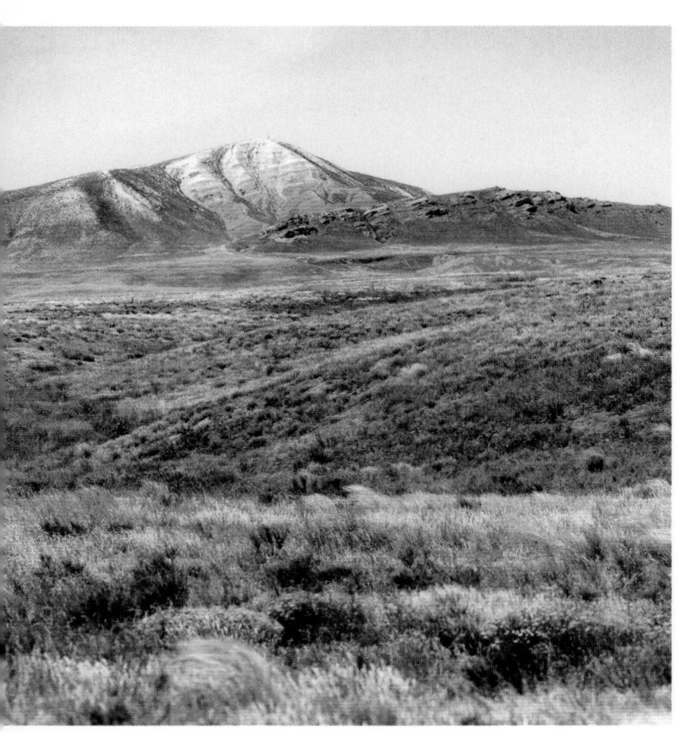

↑ Sometimes called the "Caspian steppe," such grasslands stretch from the northern shores of the Black Sea east to northern areas around the Caspian Sea.

CONSERVATION STUDIES

The grasslands of Europe, once threatened with almost complete destruction, are now receiving attention from conservation biologists. The European Union's Biodiversity Strategy, for example, aims to restore swathes of grassland among other habitats by 2030. In North America, conservationists are studying the moth faunas of the prairies, the habitat that once saw millions of bison roaming and maintaining the grassland, and some prairie restoration programs are now underway. Scientists are also examining the vast steppes of southern Europe and tropical savanna habitats, but fewer surveys are targeting the Pampas of South America or the vast Mongolian steppes, and data on their moths is scattered in obscure taxonomic monographs.

EURASIAN STEPPES

In Eurasia, grasslands covered with tufted grasses and sprinkled with wild tulips, irises, and sages, stretch from the Pontic–Caspian steppe (an area of almost 390 sq miles/1 million sq km) northward to the East European forest steppe—the land of grasses and shrubs such as *Caragana* legumes and *Prunus* (wild cherries). Further east on the Mongolian–Manchurian steppe, Stipa and Festuca grasses, among others, and wormwood (*Artemisia*) dominate the landscape. The herbs and shrubs of these vast habitats support a number of beautiful moth species, such as the Small Eggar

(*Eriogaster lanestris*) and the Small Emperor Moth (*Saturnia pavonia*). In the flower moth family Scythrididae, a large number of species have been described in the last 20 years, but many more of this and other families have yet to be studied and identified.

NORTH AMERICAN PRAIRIE MOTHS

In North America, prairies cover about 1.4 million sq miles (362.5 million sq ha) and receive about 12–20 in (300–500 mm) of rain a year. Originally, they stretched from Canada south to Indiana and Texas along the Rocky Mountains, but now have shrunk considerably under pressure from agriculture. The most common prairie species are small moths with long flight periods, that form several generations per year, and tend to feed as caterpillars on legumes. Certain species are unique to the prairie, seeking out resources only found there. They include moth species in the genus *Eucosma* (Tortricidae), numbering at least 150 species in North America, which, as caterpillars, feed on herbaceous plants such as sagebrush (*Artemisia* spp.) and compass plant (*Silphium laciniatum*).

Species within the diverse genus *Papaipema* (Noctuidae), whose larvae bore into the roots and stems of a variety of plants, derive their common names, such as the Blazing Star Borer (*Papaipema beeriana*), from the plants they attack. The Blazing Star Borer flies only in an undisturbed prairie habitat close to its host plant. There are 47 *Papaipema* species, found throughout North America in the prairies and other grassy habitats. For instance, the Rigid Sunflower Borer (*P. rigida*) feeds on several flowers such as the stiff sunflower, smooth oxeye, and golden alexander, and the larvae of the Indigo Stem Borer (*P. baptisiae*) can be found inside the stems of wild indigo and dogbane.

↖ Prairie grasses blowing in the wind in the American Midwest—its prairies are characterized by hot summers and cold winters.

↑ The Ironweed Borer Moth (*Papaipema cerussata*) is widespread in eastern North America; its common name reflects its favored host plant, New York ironweed (*Vernonia noveboracensis*).

DAPHNIS NERII

Oleander Hawk Moth

Shaped like a fighter jet

SCIENTIFIC NAME	*Daphnis nerii* (Linnaeus, 1758)
FAMILY	Sphingidae
NOTABLE FEATURES	Long proboscis, airplane shape, camouflage colors
WINGSPAN	3½ – 4¼ in (90–110 mm)
SIMILAR SPECIES	Other species of *Daphnis,* such as *D. hypothous* (India to Australia)

The beautiful Oleander Hawk Moth is fast flying but also hovers like a hummingbird above flowers, unraveling its long proboscis to sip their nectar. The moth can cover great distances, migrating from Africa as far north as Finland and breeding as far north as the Black Sea coast. To the east, its distribution extends to Japan and northern Australia.

NIGHT FEEDERS

Daphnis nerii feeds after dusk on a variety of flowers, including honeysuckles and petunias, which are fragrant at night, and is considered an important pollinator of the vulcan palm (*Brighamia*) in Hawaii. After mating, the female lays her eggs singly—frequently on the shrub oleander (*Nerium oleander*), the species' chief host plant.

APOSEMATIC CATERPILLARS

Oleander is cultivated worldwide, assisting the spread of *D. nerii,* which was introduced to control the shrub in Hawaii. As the plants are packed with toxic cardiac glycosides, the green larvae, which have a long, black, hornlike projection on their last segment, acquire a stunning coloration as they feed and develop, with a bright blue defensive eyespot on their thorax to indicate their toxicity. If consumed, they can cause vomiting in small predators, such as birds.

FAMILY DIFFERENCES

The fast-flying hawk moths of the Sphingidae family all have the same fighter-jet form, but other leaf-shaped, slower-flying hawk moths behave more like Saturniidae silk moths and do not feed. Among the feeding hawk moths, many are strictly nocturnal, pollinating only night-blooming flowers, while other smaller family members are exclusively active during the day and mimic bumblebees in size and appearance.

→ The camouflage colors of the Oleander Hawk Moth conceal it against the foliage when at rest during the day.

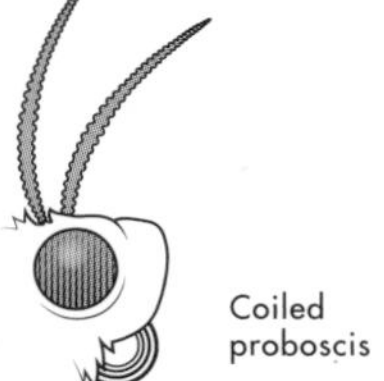

Sipping from flowers

The long proboscis gives the Oleander Hawk Moth a competitive advantage, enabling it to suck nectar from deep flowers. When at rest or flying, the moth holds the proboscis coiled, extending it in flight while hovering close to the flower.

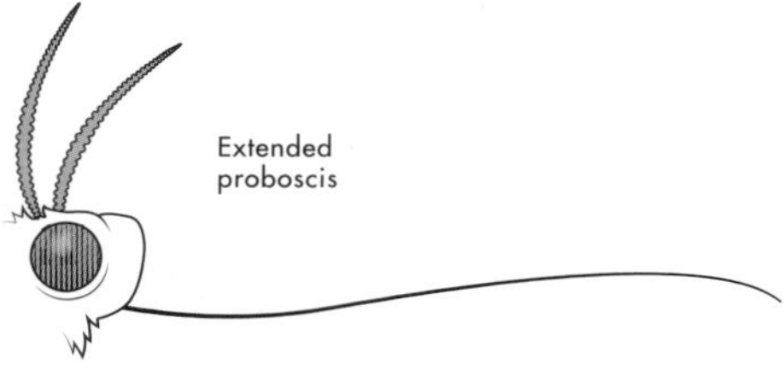

SATURNIA PAVONIA

Small Emperor Moth

Striking eyespots

SCIENTIFIC NAME	*Saturnia pavonia* (Linnaeus, 1758)
FAMILY	Saturniidae
NOTABLE FEATURES	Four eyespots, the orange hind wings of males
WINGSPAN	2⅜ – 3⅛ in (60–80 mm)
SIMILAR SPECIES	*Saturnia pavoniella* (southern Europe); *S. spini* resemble *S. pavonia* females

A widespread but never very common species, found from Scandinavia to Mongolia, the striking Small Emperor Moth occurs in open areas including grassland, heathland, moorland, and steppes, and is Britain's only resident saturniid.

MALE AND FEMALE DIFFERENCES

While both sexes have striking eyespots on all four wings, the males have orange hind wings, while the larger females are more uniformly gray. The males also fly by day, using their prominent, feathery antennae to track down females, following a trail of pheromones. The females, which lack such antennae, fly only by night, and are more sedentary and rarely attracted to light. The adults fly in spring in one generation annually and, like other saturniids, do not feed.

POLYPHAGOUS, WELL-PROTECTED LARVAE

The females lay large batches of eggs on the species' many host plants. The young caterpillars, which are black with yellow stripes, feed initially in groups and are most frequently found on shrubs such as heather (Ericaceae) and hawthorn, wild rose, wild cherries, and other Rosaceae spp. in open habitats.

When fully grown, the caterpillars change dramatically in appearance, becoming cryptic-green with numerous bright orange, yellow, or pink globular outgrowths (scoli) spread equidistantly along their length, each encircled by a black ring and covered in hollow bristles, providing a strong visual signal to potential predators. If attacked, a gland within each scolus produces an offensive, sticky, protein-rich liquid, secreted via the bristles. When mature, the larvae create a light brown, parchment-like, thin cocoon in which the pupa overwinters.

Defensive bristles

Each spheric outgrowth on the Small Emperor Moth's caterpillar not only carries spines that protect it, but also contains a gland that secretes defensive compounds.

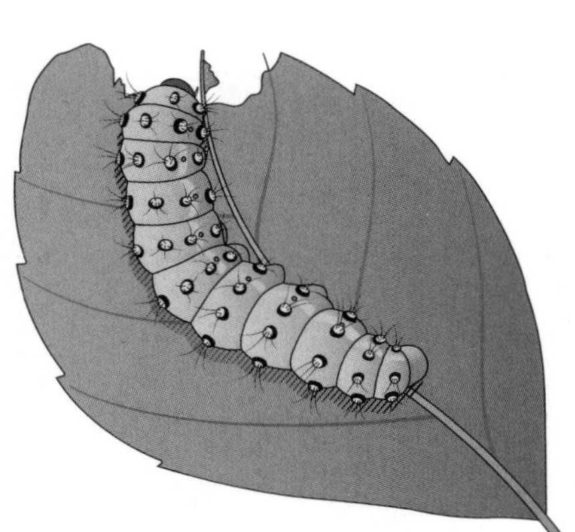

→ The Small Emperor Moth—female (top), male (bottom)—is a sexually dimorphic species (the male and female have different wing patterns).

ANANIA HORTULATA

Small Magpie

Small and delicate

SCIENTIFIC NAME	*Anania hortulata* (Linnaeus, 1758)
FAMILY	Crambidae
NOTABLE FEATURES	Triangular wings, abdomen curved dorsally
WINGSPAN	1⁵⁄₁₆–1⅛ in (24–28 mm)
SIMILAR SPECIES	*Anania shanxiensis* (sister species); the Magpie Moth (*Abraxas grossulariata*), a geometrid with a similar pattern

A familiar sight in Europe including the UK, the Small Magpie has an orangey head and thorax, and markings that can range from dark gray to almost black on its white forewings. It is possibly a more recent arrival in North America, where it occurs from Labrador to Wisconsin south to Maryland in the east, and from British Columbia to Portland, Oregon in the west. It is found in open areas wherever nettles, its preferred host plants, grow. Frequently attracted to light, this moth usually flies from May to September, but earlier or later in some areas.

OVERWINTERING PREPUPA

Like in many crambids, the green caterpillar is cryptic. When mature, it makes a tough, double cocoon—the outer part, incorporating leaf debris, protects a loose cocoon inside. Within it, the prepupal stage overwinters, resembling a shorter, stouter, and whiter version of its caterpillar, but lacking most of its functions. In spring, it sheds its skin, turns into a pupa and, within a few weeks, the adult ecloses.

PERHAPS A MIMIC

The Small Magpie can be confused with the similarly patterned, but larger Magpie Moth (*Abraxas grossulariata*), a toxic geometrid, distinguished, however, by the wavy orange markings on its forewings and wider, less shiny wings. As the majority of crambids, including most other members of the genus *Anania*, numbering more than 100, are drab-colored, the Small Magpie Moth may be an *A. grossulariata* mimic. *Anania shanxiensis* from China was recognized as a separate, but very similar, species only in 2019.

→ The Small Magpie Moth is distinguished from the larger geomtrid Magpie Moth by its triangular wings and slightly curved abdomen.

BUCKLERIA PALUDUM

European Sundew Moth

Feathery micromoth

SCIENTIFIC NAME	*Buckleria paludum* (Zeller, 1839)
FAMILY	Pterophoridae
NOTABLE FEATURES	Narrow featherlike wings
WINGSPAN	½ in (12 mm)
SIMILAR SPECIES	Other Pterophoridae, e.g. *Buckleria parvulus* (southeastern US)

Wide ranging in its Eurasian distribution, the European Sundew Moth is a tiny moth named for its larvae's unique ability to feed on leaves of its carnivorous, sundew (*Drosera*) host plants. Like other members of Pterophoridae, it is known as a plume moth for its unusually modified feathery wings which it rolls to resemble a dried blade of grass when at rest on the ground, concealing it from predators. Found in moist habitats, such as moorland and peat bogs, where its host plants grow, the moth flies close to the ground in the afternoon and is attracted to light in the evening.

ESCAPING THE TENTACLES

Its hosts, which include the round-leaved sundew (*D. rotundifolia*), secrete drops of mucilage through their trichomes—tiny, tentacle-like hairs on their leaves that trap insects attracted to the shiny, sweet fluid. The trichomes then bend toward the prey which, enveloped by the secretions, dies of asphyxiation and is absorbed by the leaves as the plant releases digestive enzymes.

Sundew Moth caterpillars, however, have evolved remarkably to overcome the potential perils of *Drosera*. The green or burgundy-colored larvae first lick the mucilage from the trichomes so that they can crawl onto a leaf and feed on it. The caterpillars, which are equipped with long hairs that may help them sense the sticky surfaces and avoid becoming trapped, also eat the trichomes, cutting them off at the base.

Mature larvae pupate head down on blades of grass without a cocoon, and the pupae are cryptic. There are usually two generations a year.

→ The European Sundew Moth surrounded by morning dew.

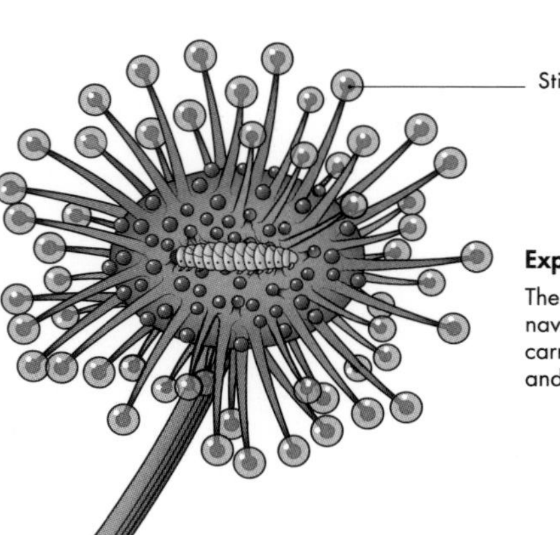

Expert maneuvers

The caterpillar of the Sundew Moth navigates its perilous way around the carnivorous sundew plant that it lives and feeds on.

EUPLAGIA QUADRIPUNCTARIA

Jersey Tiger

Distinctive tiger stripes

SCIENTIFIC NAME	*Euplagia quadripunctaria* (Poda, 1761)
FAMILY	Erebidae
NOTABLE FEATURES	Dark, striped forewings, pink or red underwings with four spots
WINGSPAN	2¹⁄₁₆–2⁹⁄₁₆ in (52–65 mm)
SIMILAR SPECIES	*Euplagia splendidior* (Middle East—e.g. Armenia, Turkey, Iraq) with smaller hind wing spots and more shimmering forewings

The Jersey Tiger takes its name from one of the Channel Islands but occurs in Europe, mainly from the Baltic states to the Mediterranean coast and east to Russia's Ural Mountains, and in Middle Eastern countries, such as Turkey and Iran.

DIURNAL AND NOCTURNAL

The adults feed on various flowers, usually by day in warmer areas, preferring Asteraceae, though often attracted to buddleia, too. In southern Britain, however, adults nectar and mate just after dark. Here, the moths, first noted at the end of the nineteenth century, are becoming more numerous and appear to be active both by day and by night, frequently attracted to light. In Britain and in similarly marginal populations in the Middle East, the species' hind wing colors can vary from bright red to orange or yellow.

OVERWINTERING LARVA

The female moth lays her eggs in batches, and the tiny larvae overwinter, emerging to feed in the spring on herbaceous plants from nettles and dandelion to comfrey, ragwort, and raspberries. When mature, they are hairy and black, with an orangey dorsal stripe, and pupate in a cocoon in leaf litter.

SEEKING SUMMER SHADE

On occasions, the Jersey Moth estivates in huge numbers on Mediterranean islands, including Rhodes, gathering in cool areas, such as under rocks and on tree trunks. Unusually, both sexes can detect the pheromones females produce, which may help them congregate. In a 2021 paper, the Jersey Tiger was listed as one of the moth species that increasingly take shelter in caves by day in response to climate warming.

→ The Jersey Tiger's boldly striped forewings conceal brightly colored hind wings.

Moth multitude

Gathering in millions on the Greek island of Rhodes, Jersey Tiger moths cluster together in cool places, such as tree trunks, to shelter from the heat.

ZYGAENA FILIPENDULAE

Six-spot Burnet

Spotted wasp mimic

SCIENTIFIC NAME	*Zygaena filipendulae* (Linnaeus, 1758)
FAMILY	Zygaenidae
NOTABLE FEATURES	Wasplike wings, clubbed antennae
WINGSPAN	1 – 1⅘ in (25–46 mm)
SIMILAR SPECIES	Other *Zygaena* species, *Z. lonicerae* and *Z. trifolii*

One of the most widespread species of burnet moths, found from Europe, Asia Minor, and the Caucasus to Syria and Lebanon, this beautiful, vivid, wasp-shaped moth has 25 subspecies. From clearings to sea cliffs and alpine meadows over 6,500 ft (2,000 m) in elevation, the day-flying species occupies a variety of grassland habitats, flying slowly a few feet above the ground during the day. The moths often sit openly on flowers such as knapweed or bird's foot trefoil (*Lotus corniculatus*), which is also the larval host plant.

CATERPILLAR HABITS

Caterpillars of the Six-spot Burnet are colored light green with rows of black spots and, like all larvae of the genus *Zygaena*—many of whom share habitats—are short, stout, and hairy. The caterpillars overwinter, then pupate in early summer inside silvery boat-shaped cocoons, attached to the stems of meadow plants.

A MOTH FULL OF CYANIDE

The moth's conspicuous, defensive coloring at all stages is derived from the cyanogenic glucosides (linamarin and lotaustralin) that the larvae obtain from the host plant, and that they, as well as the female moths, can also synthesize. The toxic compounds in the host plant likely play a nutritional role, too, as the larvae grow faster when more of them are present. The adult male transfers protective compounds to the female with his sperm during mating; the more compounds he has, the more attractive he is to the female, who can sense them by the odor of their volatile derivatives. Females also release hydrogen cyanide—probably to attract males from a greater distance.

→ Six-spot Burnet moths resting on a dandelion flower.

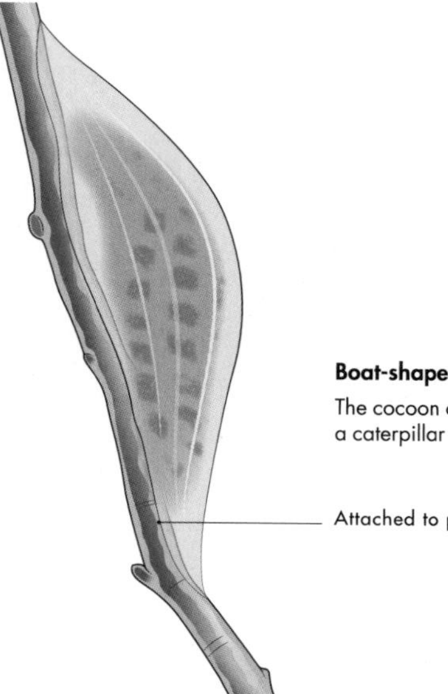

Boat-shaped cocoon
The cocoon of the Six-spot Burnet with a caterpillar ready to pupate inside.

ABANTIADES HYALINATUS

Mustard Ghost Moth

Attractive Antipodean

SCIENTIFIC NAME	*Abantiades hyalinatus* (Herrich-Schäffer, 1853)
FAMILY	Hepialidae
NOTABLE FEATURES	Large, narrow-winged, purple hind wings
WINGSPAN	4 – 4¾ in (100–120 mm)
SIMILAR SPECIES	Other *Abantiades*, such as *A. barcas*

The impressive size and invariably reddish-purple hind wings of the Mustard Ghost Moth are its characteristic features, while its forewings may vary from yellow or red to brown and are sometimes streaked with silver. The species occurs only in Australia—in Victoria, eastern New South Wales, southeastern Queensland, and Tasmania. The moths fly in February, March, and April—late summer to early winter in Australia—usually after the first rains. They do not feed and live only a few days.

ROOT BORERS

Like those of more than 30 other *Abantiades* species, the moth's grub-like, white larvae live underground where they bore into and feed on roots of plants, including black peppermint (*Eucalyptus amygdalina*) in Tasmania and bluegum eucalyptus (*E. globulus*) in plantations in Western Australia. Their heavily sclerotized head and thoracic shield, together with their strong mandibles, help them tunnel through the woody, eucalyptus roots.

FOOD FOR OTHERS

The larvae may have once supplemented the diet of Aboriginal Australians, as did at least two other *Abantiades* species on record from half a century ago, together with the better-known Witchetty Grub of the Australian desert (see page 169). The adult moths fall prey to the eastern quoll (*Dasyurus viverrinus*), an endemic Australian carnivorous marsupial and to birds, such as the black-faced monarch (*Monarcha melanopsis*).

ANCIENT, SPECTACULAR FAMILY

Mustard Ghost Moths belong to the primitive family Hepialidae and are among the world's largest moths. They are also among the most ancient, representing a family that originated more than 100 million years ago.

→ The Mustard Ghost Moth is one of the larger species of Australian hepialids.

MOTHS OF DESERTS & TUNDRA

Extreme habitats

Both desert and tundra are extreme habitats with little plant life to sustain Lepidoptera and prolonged harsh conditions, interrupted by a brief growing season. Most deserts receive less than 10 in (250 mm) rainfall a year, and moisture evaporates rapidly in the extreme heat, so only the hardiest plants and those that retain water can survive there. In tundra—the vast, treeless, far-northern regions of North America and Eurasia, together with sub-Antarctic islands—plants grow only in the short summer, when the thin layer of topsoil thaws above the permanently frozen layer (permafrost) of soil, rocks, and sand beneath it.

ADAPTED TO THE FUTURE

Moths that live in these biomes must be adapted to the extreme temperatures and to feeding on slow-growing, sometimes sparse plants. Desert and tundra moths, or very similar species, can also occur where habitats similar to these biomes exist in isolated pockets on high mountains and other very arid or cold places. Earth's temperate regions are already warmer than they were in the recent past and are likely to become even hotter in the near future, so moths that now occur in extreme habitats may be well-equipped to thrive as the climate changes.

↑ Global map of desert (red) and tundra (orange) regions.

← Greenland tundra blossoms during the summer months when temperatures can rise above 68°F (20°C) and views extend for many miles, due to the clean, dry air.

→ *Noctueliopsis aridalis*, a crambid moth found in California's Mojave Desert. Like many crambids, *N. aridalis* is colorful and feeds on flowers, but its life history is yet to be described.

POLLINATION BY MOTHS

Many desert plants bloom after dusk to avoid the heat of the day, and depend on moths for their reproduction. The fragrant flowers of the baobab tree, whose nine species grow in Africa, Madagascar, and Australia, open at dusk, attracting hawk moths, such as *Coelonia* spp. Hawk moths also visit the desert rose of Africa and the Middle East which, like the baobab tree, has a swollen trunk to conserve moisture in times of drought. Many nectar-feeding moths, together with bats, visit blooming agave at night and are important pollinators in the desert. Most of the flowers pollinated by hawk moths have long tubular corollas into which the proboscis is inserted, but there are exceptions: the nocturnal Five-spotted Hawk Moth (*Manduca quinquemaculata*) nectars on both exotic and native desert garden cacti, such as *Peniocereus greggii*, found in the Sonoran Desert from Arizona to northern Mexico. While cactus blooms visited by hummingbirds are usually brightly colored and open by day, those pollinated by moths are white and open toward nighttime.

← ↑ Desert Rose trees, such as these on Socotra island, east of the Horn of Africa, are pollinated by hawk moths. The Convolvulus Hawk Moth (*Agrius convolvuli*) occurs in the region.

Subtropical desert moth communities

The hottest and driest places on Earth are located away from the Equator and are mostly associated with subtropical continental climates. They include the African Sahara and Australia's Great Victoria Desert, the Mojave and Sonoran deserts of North America, Lut Desert in Iran, and many others, collectively accounting for 26–35 percent of dry land.

STRIKE WHILE IT RAINS

The intense heat of the summer is relieved by cooler but still dry winter temperatures, mostly above freezing point. Whatever rain falls tends to come in quick bursts, and sometimes no rainfall reaches the ground at all. Moths and many other creatures are active by night when temperatures drop. While the line between grasslands and deserts is often blurred, with semideserts frequently forming transitional habitats, what largely defines moths that are found here is their ability to form associations with drought-adapted plant species, such as cacti, yuccas, palms, and some hardy trees and shrubs. Moths from surrounding areas may also seasonally migrate into deserts after the rains to take advantage of short-lived plant growth.

DESERT MOTH DIVERSITY

The more biologists explore extreme and less-studied habitats across the globe, the more they begin to discover their unique diversity. In 2006, a survey initiated in the White Sands National Park in New Mexico, at the northern limits of the Chihuahuan Desert, yielded over 450 species of moths, including 19 that were new to science—many of them probably unique. Some of the associations between desert plants and moths are also ancient, dating back millions of years to Cretaceous-Jurassic times, as in the case of the primitive welwitschia (*Welwitschia mirabilis*) of southern Africa's Namib desert, pollinated by pyralid and geometrid moths. Certain leaf-mining micromoths also feed on this relict, sometimes described as a "living fossil," supporting the hypothesis that moths were present and important in plant biology even before the emergence of the earliest flowering plants.

PLANT COMMUNITIES

Plant communities in the deserts are highly distinctive. In the Mojave Desert, for example, dominant species include brittlebush, desert holly, Joshua tree, and creosote bush. Many specialist moths have evolved to utilize these plant resources, such as the Creosote Moth (*Digrammia colorata*), which feeds, as its name suggests, exclusively on creosote bushes; there are also cactus, yucca, and palm specialists. In Australia, the vast Great Victoria Desert habitat that stretches for more than 163,000 sq miles (42 million ha) is characterized by certain drought-resistant eucalyptus and acacia species. In 2017, a bioblitz (intense survey of various life forms) revealed 86 species of Lepidoptera, some previously unknown, feeding and flying in this harsh habitat.

↑ A caterpillar of the White-lined Sphinx (*Hyles lineata*) in the Colorado Desert. The larvae are highly variable in appearance and feed on a wide variety of desert plants.

← *Welwitschia mirabilis* found in southwest Africa is the only living species of its ancient Welwitschiaceae family and is associated with moths of an equally ancient lineage.

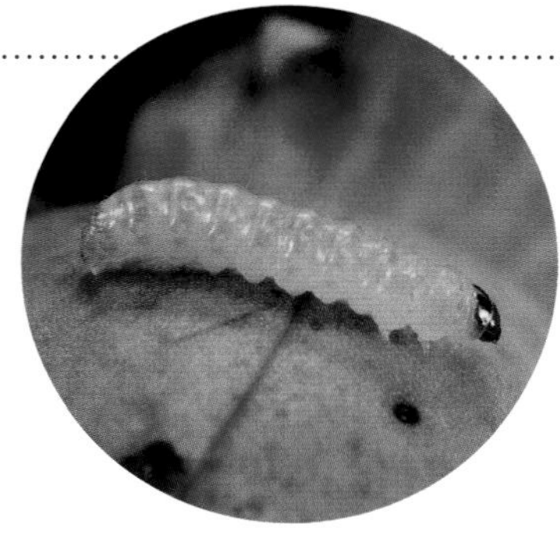

← Larva of the Pink Bollworm moth (*Pectinophora gossypiella*). While native to Australasia, the species is now found wherever cotton or its other host plants grow.

MOTHS ON COTTON

The Pink Bollworm (*Pectinophora gossypiella*) and the Queensland Pink Bollworm (*P. scutigera*) are native to Asia and Australia. These moths are well adapted to surviving dry, hot conditions. The adults mate early in the morning, around 3 a.m., after females attract males with their pheromones. Even though the female can lay eggs anywhere on the plant, it prefers to do so on the fruit, referred to as cotton "bolls." Several bollworms can develop within a single boll, then they crawl down the plant and pupate in soil. If conditions become unfavorable (too hot or too cold), caterpillars can enter a diapause before pupation occurs. Adults can disperse over great distances in the desert in search of cotton plants, and if they find a field of cultivated cotton, their population can explode. The moths can also develop on other mallows (Malvaceae), such as *Hibiscus* spp., okra, and portia tea plants, and sometimes prefer them to cotton.

Moths on cacti

Moth larvae feed on the fleshy tissues of many of the 1,700 cacti species known to science. In the Americas, five species of snout moths in the *Cactoblastis* genus burrow inside cactus stems to protect themselves from the desert heat, as do many other moth genera here and elsewhere.

EXTERNAL FEEDERS

There are a few caterpillars that feed externally on cacti: among them is the beautiful Staghorn Cholla Moth (*Euscirrhopterus cosyra*) of the southwestern United States and northern Mexico that favors chollas (*Cylindropuntia* spp.)—cacti that resemble prickly pears but are taller and thinner. Its orange, black-striped caterpillars scrape the surface and consume new growth on the chollas and sometimes other cacti, such as prickly pear and saguaro, the towering cactus that dominates the Sonoran Desert. The Variegated Cutworm (*Peridroma saucia*), Beet Armyworm (*Spodoptera exigua*), Granulate Cutworm (*Feltia subterranea*), and Corn Earworm (*Helicoverpa zea*) all attack young growth on saguaro before the cactus reaches its fifth year. These noctuids are frequently migratory—breeding in Arizona and other southern states in the winter and spreading northward during the summer all the way to Canada, feeding externally on a variety of hosts, including agricultural crops.

MUTUALLY BENEFICIAL

The Senita Moth (*Upiga virescens*) of the Sonoran Desert, the only species of its genus, has a mutually beneficial relationship with the senita cactus (*Lophocereus schottii*). This nocturnal moth has special adaptations to transfer pollen from one flower to another and is responsible for about 75 percent of all successful pollination of the senita. The female lays a single egg on each plant, and the larva bores into the flower and feeds on developing fruit and seeds. As Senita larvae consume only about 30 percent of the seeds, the moths have an overall beneficial effect on the cactus, as it assures pollination: without the moth, fewer senita cacti in a community would reproduce at all.

← A mature caterpillar of the Staghorn Cholla Moth (*Euscirrhopterus cosyra*) feeds on its cholla (*Cylindropuntia* sp.) host plant.

THE CACTOBLASTIS MOTH

The South American Cactus Moth (*Cactoblastis cactorum*), one of five *Cactoblastis* species, is considered a hero in Australia, but a pest elsewhere. As the name suggests, its caterpillars feed inside the fleshy stems of the prickly pear cactus (*Opuntia* spp.), causing considerable damage. The moth was introduced in Australia during the first half of the twentieth century, when the non-native prickly pears overran vast swathes of land. Within a few years, *C. cactorum* brought the prickly pear under control in one of the most successful biological control programs ever. While the Australians put up a monument to the moth to honor its service to Queensland, elsewhere its story is much less positive. Unlike most other cactus-feeding moths, which have restricted diets, the South American Cactus Moth not only kills prickly pears but also other types of cacti. When introduced in South Africa and the Caribbean, it attacked many native varieties, endangering them and their associated wildlife. In Mexico, Texas, and Argentina, where prickly pears are an important crop (used to produce cheeses, flours, nectars, and fruits in syrup), there are fears that *C. cactorum* may become invasive and seriously affect the livelihoods of many people.

The South American Cactus Moth example is one of many human introductions of potentially "useful" alien plants that can go very wrong. Within its native South American range, the moth is naturally controlled by herbivores, parasitoids, germs, and predators in associations that have evolved over millions of years. Introduced outside their range, a species can spiral out of control, breeding exponentially without the other species to keep its numbers in check.

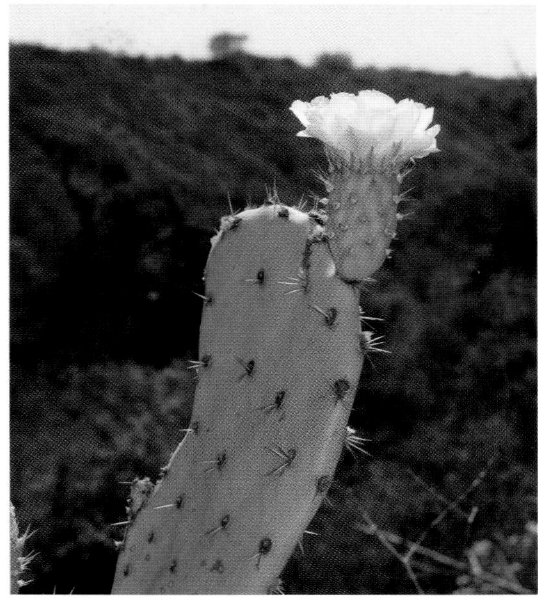

↗ A flowering prickly pear (*Opuntia*) cactus; *Opuntia* species are ecologically and economically important in warmer parts of the New World but can be invasive exotics elsewhere.

→ The South American Cactus Moth (*Cactoblastis cactorum*) has been introduced in a number of regions to control invasive prickly pear cacti. The caterpillars feed in groups internally, gutting the fleshy stem of the cactus.

Moths on yuccas

Among the most important desert creatures, are the tiny yucca moths of the genus *Tegeticula,* whose 20 North American species pollinate and then feed on the seeds of *Yucca* spp. that grow in desert regions and also grasslands and prairies.

COOPERATIVE RELATIONSHIPS

Like the Senita Moth (*Upiga virescens*), yucca moths have a symbiotic relationship with their hosts; the cooperation of the two organisms is essential for the survival of both. Both males and females are attracted by the aromatic scent of the flower and also mate within it. The female collects pollen from anthers (where pollen is produced), then rolls it into a ball and stuffs it into the stigma (where pollen germinates) of the same flower. To adapt to their pollination role, the moths developed especially long, curled maxillary palpi (sensory appendages) that facilitate the insertion of pollen into the stigmatic orifice of the yucca. While this enables seeds to develop, the moths have an ulterior motive: females also penetrate the ovule (where seeds germinate) with their sharpened ovipositor and lay a few eggs. The hatchlings then feed on ripening seeds, but only eat a portion of them, leaving enough available for dispersal to propagate the host plant.

Unlike cutworms that feed on saguaro cactus, but can as easily feed on other plants, *Tegeticula* feed exclusively on yuccas. For example, around San Diego, California, *T. maculata* pollinates and feeds on *Yucca whipplei*, while outside of Las Vegas in the Mojave Desert, *T. yuccasella* is associated with the Mojave yucca (*Yucca schidigera*).

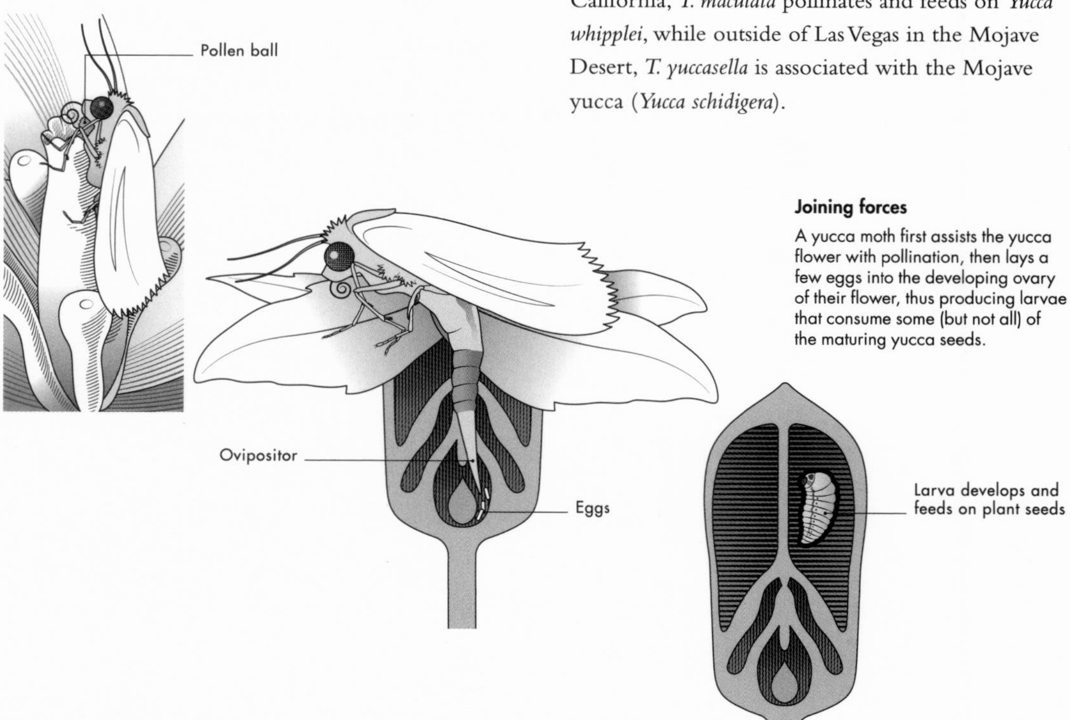

Joining forces

A yucca moth first assists the yucca flower with pollination, then lays a few eggs into the developing ovary of their flower, thus producing larvae that consume some (but not all) of the maturing yucca seeds.

JOSHUA TREE

Tegeticula synthetica, described by British-born entomologist and artist Charles Valentine Riley in the nineteenth century, feeds on and pollinates only the iconic Joshua tree (*Yucca brevifolia*) most commonly found in the Mojave Desert. In 2003, researchers described a second, slightly smaller moth species—*T. antithetica*—associated with the Joshua tree. This seemed unusual at first as yucca moths rarely, if ever, compete on the same host plant. It transpired, however, that *T. antithetica* occurs mainly in eastern parts of Mojave Desert scrub at elevations of around 3,000–5,500 ft (900–1,700 m), together with the *Y. brevifolia* variant, known as *jaegeriana*—the Eastern Joshua tree. Genetic differences between moth populations of the two species suggest that a split occurred around 10 million years ago, as the woodland savannas were transforming into semideserts in this area, and the Joshua tree range was expanding and contracting with fluctuations in climate. So, the appearance of a new species can be, as is frequently the case in moths and other animals, a result of prolonged geographic isolation of two populations, reinforced by adaptations to different environments and slightly different host plants.

↑ A Joshua Tree (*Yucca brevifolia*) in the Mojave Desert. According to legend, Mormon immigrants gave it its biblical name.

Moths on palms

Exclusive caterpillar–host plant associations similar to those between cactus moths and cacti are also found among many palm-feeding species that occur both in deserts and desertlike habitats, where sandy soil does not hold much moisture.

TUNNELING CATERPILLARS

In sandy habitats along the southeastern coast of the United States, Cabbage Palm Caterpillars (*Litoprosopus futilis*) tunnel in the stalks of palms and palmettos. Larvae of their localized relative, *L. bajaensis*, feed on the Mexican blue palm (*Brahea armata*) in an oasis in the northern desert of the Baja California. Although they damage up to 70 percent of fresh shoots on their tall host plants that dominate this harsh landscape of granite rocks, the sweet secretions oozing from the holes made by the caterpillars attract other creatures that feed on it—an example of a moth–plant interaction that benefits the larger ecological community of insects and birds.

NEW VORACIOUS SPECIES

Scientific knowledge of palm-associated moths is far from complete: in 2021, research revealed two new species within the small family Pterolonchidae—*Homaledra howardi* and *H. knudsoni*. The first species was described from specimens found in Florida and the Dominican Republic, while *H. knudsoni*, named after Texan moth explorer Edward Knudson, was based on specimens collected in Texas, Florida, and Mexico. These two species of palm leaf skeletonizers not only strip the tough leaves of their host plant down to the veins, but also build elaborate frass tunnels, and can cause significant damage to the palms.

← Palm Leaf Skeletonizers (*Homaledra sabalella*) scrape the surface of palm leaves while hiding inside the tunnels, seen here, made of silk, frass, and debris.

↑ The date palm that now grows all over the world, is frequently attacked by Lesser Date Moth (*Batrachedra amydraula*) caterpillars that tunnel through unripe dates.

↑ The Palm Moth (*Paysandisia archon*) is a beautiful, large castniid moth native to southern South America. In 2001, it was first found in southern Europe and has spread since, attacking cultivated palms.

Wood borers

A few moth species inhabit even the hottest deserts on the planet, such as the Lut Desert in Iran, with the hottest temperature on record of 159°F (70.7°C). Among them are carpenter moths (Cossidae), which have grub-like larvae that burrow through the trunks, stems, and roots of woody plants where the temperature is cooler.

DESERT SPECIALISTS

Carpenter moths are well adapted to desert conditions, and many seem to prefer them to other biomes. A 2013 analysis of all carpenter moths found to date in the deserts of northern Africa and Eurasia, north of the Himalayas, showed that more than 100 species—accounting for 38 percent of all carpenter moths—fly in this vast biogeographic region. Four cossid species occur in the Sahara Desert. Of these, the widespread desert specialist *Holcocerus holosericeus* is striking for its snow-white forewings, thorax, and abdomen. The three others, which have marbled gray-brown wings more typical of cossids, are *Eremocossus vaulogeri*, also found in Israel and other Middle Eastern countries; *Paropta paradoxa*, which can feed on drought-tolerant Egyptian acacia (*Vachellia nilotica*) but elsewhere is a pest on grapes; and *Isoceras kruegeri*, known only from Libya.

Two recently described genera of cossids, *Chingizid* and *Kerzhnerocossus*, are found exclusively in Mongolia's huge Gobi Desert. The survey that uncovered them again indicates that analyzing local moth populations in isolated areas and studying their genitalic structures is likely to reveal more desert cossid species. However, much more research is needed to understand the true diversity of such deserts.

WITCHETTY GRUBS

Australia has almost 100 known carpenter moth species. One of the most important is *Endoxyla leucomochla*, whose larvae—known as witchetty grubs—feed on sap from the roots of the witchetty bush (*Acacia kempeana*) and possibly other desert plants. The grubs build underground chambers and have long been an important desert food for Aboriginal Australians who dig them up and eat them raw or lightly cooked. In a 1952 account of the moths and their use by the Aboriginal people around a tiny desert settlement of Ooldea in the south of the country, Australian anthropologist and entomologist Norman Tindale reported that the cooked larvae tasted more like pork, while raw larvae had a creamy, buttery flavor. As the grubs are rich in fat and protein, Tindale considered them a highly nutritious food for children and useful for weaning infants. The moths of this species are large, up to 6¾ in (170 mm) in wingspan and take two or more years to develop. Like other cossids, they do not feed as adults, but have a high reproduction rate.

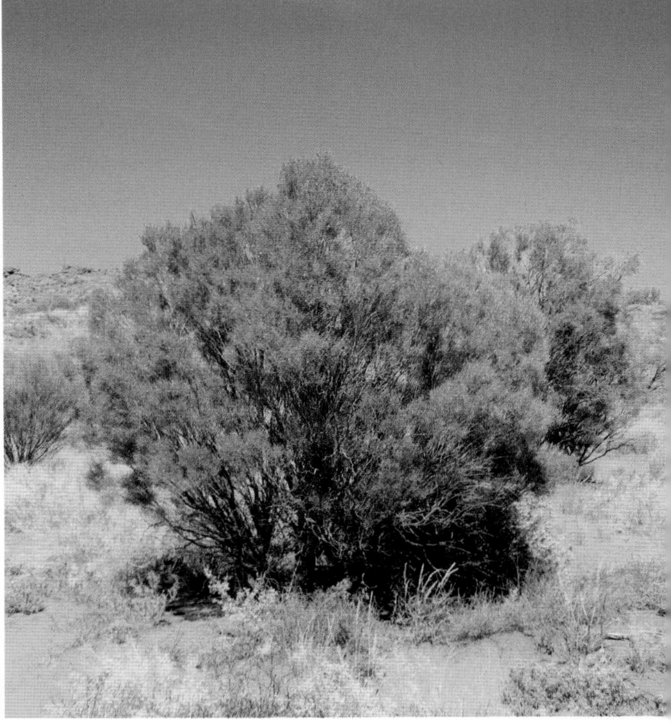

↗ In desert areas of Australia, many of the larger Lepidoptera larvae, including those of the cossid *Endoxyla leucomochla* (wichetty grubs), have been used as traditional Aboriginal foods, especially in the past.

→ *Acacia kempeana*, also known as the witchetty bush, is a desert plant on whose roots *Endoxyla leucomochla* larvae feed, hence their common name "witchetty grubs."

← Carpenter moths are relatively rare, but *Paropta paradoxa* is found throughout the Middle East, developing within the bark or stems of a variety of species, including the Egyptian acacia tree and grapevines.

Moths on mesquite

Around the world, there are 26 species of distinctive, drought-tolerant mesquite, carob, and other similar trees in the genus *Prosopis*, which are a feature of many semidesert and desert habitats. They are frequently host plants of Lepidoptera, including a variety of silk moths, from striking *Automeris* eyed moths on *Prosopis alba* and *P. nigra* in South America, to buck moths (*Hemileuca*) and Hubbard's Silk Moth (*Syssphinx hubbardi*) on mesquite in the southwestern United States and Mexico.

THE INVADER'S CONQUEROR

While mesquite is prolific, invasive, and highly capable of adapting to harsh conditions, some moths can badly damage the plant. Among major moth defoliators of mesquite are the larvae of the finely-patterned erebid moth, the Indomitable Melipotis (*Melipotis indomita*), also known as the Mesquite Cutworm. Females lay around 750 eggs, and six weeks later the next generation of moths emerges. Caterpillar density can be quite high: in one instance 86 larvae and pupae were gathered from a 1.1 sq yd (1 sq m) area under mesquite trees. The moths can be found across the United States, frequently as migrants unable to establish for long, as they cannot survive in freezing conditions.

WEB WEAVERS

The honey mesquite is also favored by gelechiid Mesquite Webworm Moths (*Friseria cockerelli*), found throughout California, Nevada, and Texas. As their common name suggests, the caterpillars make webs in the forks of mesquite; around 100 larvae may live together in a nest, where they also pupate and diapause.

← The Hubbard's Silk Moth (*Syssphinx hubbardi*) caterpillar feeds on mesquite and acacia.

→ A black-tailed gnatcatcher (*Polioptila melanura*)—a small, insectivorous bird native to the Sonora Desert—catches a caterpillar on a mesquite branch in Arizona.

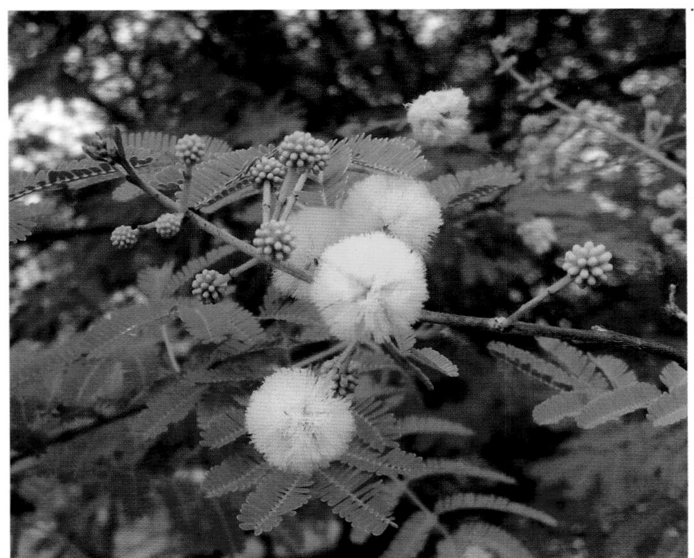

INVASIVE MESQUITE

At least 12 species of mesquite, native to the southwestern United States, Mexico, and South America have become established in parts of Africa, Asia, and Australia, frequently showing great drought tolerance. However, one species, the South American thorny mesquite (*Prosopis juliflora*), has become invasive on all three continents. In Australia, the leaf-tying moth (*Evippe* spp.) was introduced as a biocontrol measure in 1998 and appears to be successful.

← While it has a variety of uses in the New World, thorny mesquite (*Prosopis juliflora*) is invasive elsewhere.

BUCK MOTHS

In the deserts of the southwestern United States, large, beautiful, butterfly-like buck moths (*Hemileuca* spp.) are testimony to moths' ability to utilize a variety of ecological niches in different types of deserts and semideserts. Buck moth males, which got their name from their habit of flying during the opening of the deer-hunting season, often search out females by day in and around the Sonoran and Mojave deserts. As caterpillars, these moths feed on plants common in the habitat they occupy, from buckwheat to horsebrush and desert almonds, collecting all the nutrients they will need as adults.

Sixteen of 24 species of buck moths are found in semidesert habitats, chaparral scrublands, and Great Basin grasslands of the American Southwest and Mexico—in each habitat with slightly unique flora. For example, the snow-white Burns' Buck Moth (*H. burnisi*), perhaps one of the most striking moths found in and around California, feeds on an extremely drought-tolerant legume—the Mojave indigo bush (*Psorothamnus arborescens*).

Just above California's Mono Lake, which formed at an elevation of 6,383 ft (1,946 m) less than a million years ago as a result of violent volcanic activity, is a sun-scorched habitat dominated by sagebrush that American writer Mark Twain described as "lifeless, treeless, hideous desert ... the loneliest place on earth." Here, males of the Western Sheep Moth (*H. eglanterina*), colored in shades of pink and yellow, etched with black lines and spots, battle against the wind in search of sedentary females. After hatching from overwintering egg batches that encircle twigs of their host plant, their black caterpillars, which later develop stripes and orange tufts, feed on the shrubs' fresh growth.

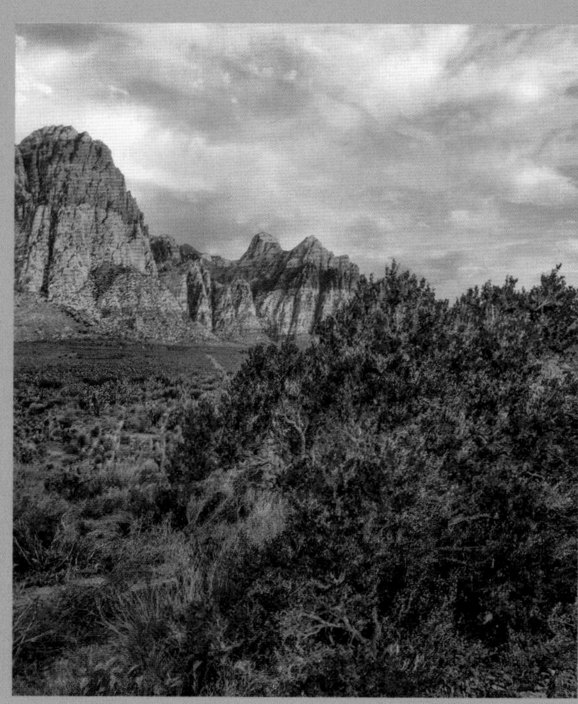

↗ The Electra Buck Moth (*Hemileuca electra mojavensis*) in California's Mojave Desert.

→ The Mojave Indigo Bush (*Psorothamnus arborescens*), native to the Mojave, Colorado, and Sonora deserts, is one of various host plants used by the buck moth (*Hemileuca* sp.).

→→ The Western Sheep Moth (*Hemileuca eglanterina*) is a day-flying saturniid found in sagebrush-covered a reas east of the Sierra Nevada.

Moths of the tundra

The tundra biome includes numerous relatively barren habitats in Alaska, northern Canada, and northern Eurasia, with faunas sometimes shared between the two continents. The moth diversity of tundra has only recently been studied, but research from Scandinavia, Russia, and Alaska already suggests that it consists of both widespread moths that fly elsewhere (some migrate only seasonally into tundra), and also unique species found nowhere else. Tundra biomes with unique moth fauna are also found in high mountains around the world, and in the sub-Antarctic region.

STARK LANDSCAPE

Tundra is relatively barren land, characterized by low-growing herbs and shrubs, such as the Labrador tea (*Rhododendron tomentosum*), and *Vaccinium* spp., such as lingonberry and bog blueberry. Many tundra tree species are dwarfed, so, for example, willows that may grow tall in temperate regions are represented here by a number of species that look like small shrubs.

WOOLLY BEARS

During the brief summer, moth caterpillars, such as woolly bears (named for their hairy appearance), crawl around tundra looking for food and prefer deciduous shrubs, such as dwarf birches and the pincushion plant (*Diapensia lapponica*). These caterpillars, from the arctiine subfamily of Erebidae known as tiger moths, are well adapted to life in the tundra. For example, the Arctic Woolly Bear Moth (*Gynaephora groenlandica*),

← The Common Heath Moth (*Ematurga atomaria*) feeding on marsh Labrador tea flowers. Following the distribution of its freeze-tolerant heather host plants, this geometrid is widespread in the harsh open habitats of northern Eurasia.

← The Williams' Tiger Moth (*Apantesis williamsii*) can develop in harsh habitats of far northern areas of North America.

↓ The overwintering caterpillars of the Ruby Tiger (*Phragmatobia fuliginosa*) moth, like many other woolly bears, are cold tolerant and polyphagous, which allows this successful species to thrive in territories of the far north.

discovered in Greenland in 1832 during an Arctic expedition led by Scottish explorer John Ross, takes up to seven years—one larval instar per year—to reach maturity and pupate. The male moths bask in the sun, warming themselves before flying to search out sedentary females, and the larvae also need the sun's warmth to raise their temperature and enable them to function. The species is also found on Wrangel Island in Russia and in Canada.

The Labrador Tiger Moth (*Apantesis quenseli*) is found in both Eurasia and North America, in places such as Alaska and Labrador, and forms isolated populations in the alpine tundra of high mountains. In these cold, harsh regions, the moth flies by day. The Polar Tiger Moth (*Arctia subnebulosa*) is widely distributed in the North American tundra, and the closely related Arctic Tiger Moth (*A. tundrana*), has been recorded in 35 different localities throughout northern Eurasia in 2021. The markings of the large, attractive adults in these sister species vary slightly in different geographic areas. Though widespread, *A. tundrana* was first described only in 1990, but has been studied extensively since in the Russian arctic tundra. According to a 2015 study, its woolly bear caterpillars are heavily parasitized by a braconid *Meteorus* wasp that destroys 90 percent of them.

Tundra moth communities

In the summer of 2012, a survey of moths conducted by an international group of scientists in low-shrub tundra habitat of northern Russia's Nenets region, found 29 Lepidoptera species. Some, such as the gelechiid *Gnorimoschema vastificum*, were previously known only from North America. *Greya variabilis*, a prodoxid moth that occurs in Alaska, was also found in Nenets tundra. A tortricid, *Eucosma ommatoptera*, was previously known only in the Far East and Japan. The survey also revealed that the ranges of the crambid *Udea uralica* and the geometrid *Xanthorhoe uralensis*, thought to be limited to the polar Ural mountains, were much greater than previously known.

← Several species of plume moths, such as this Yarrow Plume (*Gillmeria pallidactyla*), can develop in colder northern territories, overwintering as pupae in the roots of herbaceous plants.

→ *Tinagma dryadis*, representatives of the small family of Douglas moths (Douglasiidae), mating on mountain avens (*Dryas octopetala*) on the Aland Islands, Finland.

WIDESPREAD SPECIES

Common species found in the Nenets survey included a very widespread plume moth *Platyptilia calodactyla*, a species from temperate Europe also recorded from Iran, and the Common Clothes Moth (*Tineola bisselliella*)—possibly introduced by humans. *Apotomis frigidana*, a tortricid moth known from the Yukon region of western Canada, proved the most abundant species in the stony, shrubby tundra. Here, it was feeding on cushion plants known as white dryad (*Dryas octopetala*)—the emblem of Canada's Northwest Territories—which blooms briefly in spring. Further discoveries included three *Phiaris* spp., whose larvae are known to feed on tundra blueberries and other *Vaccinium* plants.

ARCTIC MIGRANTS

Moths can migrate long distances, sometimes carried by wind. In 2020, Larch Bud Moths (*Zeiraphera griseana*) were found in large numbers on Wiese Island in the Arctic Ocean in late summer, although the island has no suitable breeding habitat. The moth's sudden abundance can be linked to an outbreak (an explosive increase in numbers) in its native boreal forest more than 1,000 miles (1,600 km) to the south. As the Earth's climate warms, more moth species may gravitate toward the Poles.

SUB-ANTARCTIC COMMUNITIES

Tundra communities, though much less extensive, are also found in the sub-Antarctic region. Marion Island, whose tundra differs from other sub-Arctic areas because of the extensive manure created by seabirds and seals, is home to the Marion Flightless Moth (*Pringleophaga marioni*), which takes five years to develop on a diet of detritus, such as dead plant matter. Its adults have greatly reduced wings, while its larvae, known as sub-Antarctic caterpillars, are quite common and contribute significantly to the diet of other species on the island.

ALPINE TUNDRA MOTHS

Like the moths of the northern tundra biome, moths at high elevations in alpine tundra habitats can be unique or may represent a widespread species that migrates to the mountains for varying reasons.

A survey of nocturnal moths in Colorado at an elevation of around 12,500 ft (3,800 m), found 48 larger moth species and numerous micromoths. Among the larger moths, were several cutworms (*Euxoa* spp., *Peridroma saucia*, *Polia rogenhoferi*, *Anarta farnhami*), and a number of other noctuids, such as the Mountain Beauty (*Syngrapha ignea*). The American Lappet Moth (*Phyllodesma americana*), which resembles a dry oak leaf, was no surprise as this species commonly occurs all the way north to Yukon; the moth is well adapted to cold climates and its caterpillars feed on a variety of plants.

Less expected was *Entephria lagganata* (Geometridae), previously known only in Canada. Such cases of disjunct distribution, where a montane population is geographically separated from the main population farther north, may result from moths and their host plants favoring cooler, higher mountain habitats during periods of climate warming.

↑ Alpine azalea (*Kalmia procumbens*) is a northern plant of North America that seldom grows farther south than mountains in the states of Maine and New York in the east, and Washington in the west.

↖ The American Lappet Moth (*Phyllodesma americana*) is so polyphagous and cold tolerant that it can inhabit not only alpine tundra, but also the northern territories of British Columbia and Yukon.

THE GREAT MIGRATORS: WHITE-LINED SPHINX VS. STRIPED HAWK MOTH

While some migratory species, such as cutworms, may seek cool hideouts at high elevations, other widespread and fast-flying moths remain active in a great variety of habitats, including deserts, tundra, and high mountain tops. Their occurrence is not only seasonal but also varies greatly from year to year, depending on the weather and other factors. Many hawk moths migrate, but the best examples are the White-lined Sphinx (*Hyles lineata*) and the Striped Hawk Moth (*H. livornica*)—a New World and Old World species which, until recent DNA studies proved otherwise, were thought to be the same.

The White-lined Sphinx, whose range extends from South America to British Columbia, can be found sporadically in the alpine tundra of Colorado, visiting flowers such as valerian, mountain buckwheat, and others. It also migrates into deserts during springtime, where its caterpillars can be numerous, munching on a variety of desert plants such as sand verbena (*Abronia umbellata*) and toothed spurge (*Euphorbia dentata*). Cahuilla and Tohono O'odham Native American peoples in California used to collect these caterpillars, roasting them for immediate use or drying and storing them. These highly variable caterpillars—sometimes green with eyespots, and sometimes almost entirely black—are often seen crawling around, searching for a plant to defoliate. Similar caterpillars of the Striped Hawk Moth, which occurs from South Africa to Eurasia, were recently reported in huge numbers in the Sinai desert after rain, feeding on a variety of desert plants, while the moths took nectar from *Iphiona* flowers.

↓ The White-lined Sphinx Moth (*Hyles lineata*) is a widespread species, sometimes found in extreme habitats thanks to its ability to migrate and utilize a variety of hosts as a caterpillar.

XANTHOTHRIX RANUNCULI

Xanthothrix ranunculi

Minuscule Mojave Desert moths

SCIENTIFIC NAME	*Xanthothrix ranunculi* (Hy. Edwards, 1878)
FAMILY	Noctuidae
NOTABLE FEATURES	Golden or silvery, uniformly colored forewings, occasionally with a white discal spot; black hind wing scales with a hint of gold
WINGSPAN	$^{5}/_{16}$ – $^{7}/_{16}$ *in* (8–11 mm)
SIMILAR SPECIES	None, except *X. ranunculi albipuncta*, sometimes considered a separate species

Found in the Mojave Desert, where it is thought to be endemic, *Xanthothrix ranunculi* feeds on nectar from the yellow flowers of the Douglas' tickseed (*Coreopsis douglasii*), which sprouts annually in California, growing in small patches on southern-facing slopes. The coloring of the adults' wings and highly patterned larvae camouflages them against the flowers.

OBSERVATIONS IN THE DESERT

The species life history has only once been fully described. The renowned American entomologist John Henry Comstock (1849–1931) observed and collected several *X. ranunculi* females after heavy rains over the Mojave Desert and encouraged them to oviposit on their *C. douglasii* host plant. He subsequently recorded the life cycle of this uncommon and highly local moth species, although the account was published in 1941, after his death, coauthored by Christopher Henne.

CRYPTIC LIFE HISTORY

According to Comstock, the smooth, white, oval eggs, about 0.5 mm in length, are laid among petals and seeds, which the caterpillars feed on. At first, the young caterpillars are yellow with a black head and become more patterned, with a notched middorsal stripe down the back, at later instars. The mature larvae are stout and cylindrical, a creamy yellow in color, with orange-brown bands and black spiracles, and around ½ in (13 mm) in length.

Xanthothrix ranunculi forms one generation a year at most, but, according to Comstock's assessment, may be capable of spending more than a year in the pupal stage during a drought, when conditions for reproduction are unfavorable.

→ *Xanthothrix ranunculi* moths feed, mate, and lay eggs on the flower of their host plant, the Douglas' tickseed (*Coreopsis douglasii*).

SYSSPHINX HUBBARDI

Hubbard's Silk Moth

Cryptic and striking

SCIENTIFIC NAME	*Syssphinx hubbardi* (Dyar, 1902)
FAMILY	Saturniidae
NOTABLE FEATURES	Distinctive mouse-gray forewings concealing pink hind wings with an eyespot
WINGSPAN	2⅝ – 3 1/16 in (66–77 mm)
SIMILAR SPECIES	*Syssphinx heiligbrodti*, a paler species found from south and central Texas to Colorado, overlapping with *S. hubbardi* locally

Found from the Sonoran and Mojave Deserts (southern Texas to southern California) in the United States to southern Mexico, slow-flying *Syssphinx hubbardi* moths of both sexes are similarly colored, but the male is smaller than the female. The subdued, cryptic gray of their upper forewings helps to conceal them, but, if disturbed, they will startle predators by revealing striking, pink hind wings marked with a single black eyespot and similarly colored ventral forewings when threatened. In the Arizona desert, the pallid bat is known to feed on the moths, which are quite common from May to October.

DEFENDED BUT CRYPTIC CATERPILLARS

The Sonoran Desert and the semidesert of Trans-Pecos and Staked Plains in Texas are home to the caterpillars of Hubbard's Silk Moth. Among the dry grasses, scrub oak, and other drought-tolerant plants of these areas is the caterpillars' preferred host—honey mesquite (*Prosopis glandulosa*), a common plant, whose flowers are favored by butterflies and whose seed-pod lining local people use to produce a protein- and carbohydrate-rich flour.

The larvae hatch with eight formidable thoracic projections, covering about half of their length, and with a posterior horn, more typical of sphinx moths caterpillars. As they develop, they continue to be well defended by the horn and by a comb of sharp spines along the length of their back. While their striation camouflages them against the narrow leaves of their host plants, the candy stripe of red and white along their side can appear deceptively aposematic.

→ The warning coloration of Hubbard's Silk Moth is just a bluff as it is not toxic, but its eyespots that simulate the eyes of raptors or snakes are enough to frighten off small birds.

Defensive spikes

Despite its seemingly bright coloration, a caterpillar of Hubbard's Silk Moth is cryptic among leaves of mesquite and it is not defended by any toxins, though its spikes may make it difficult for a small bird to swallow.

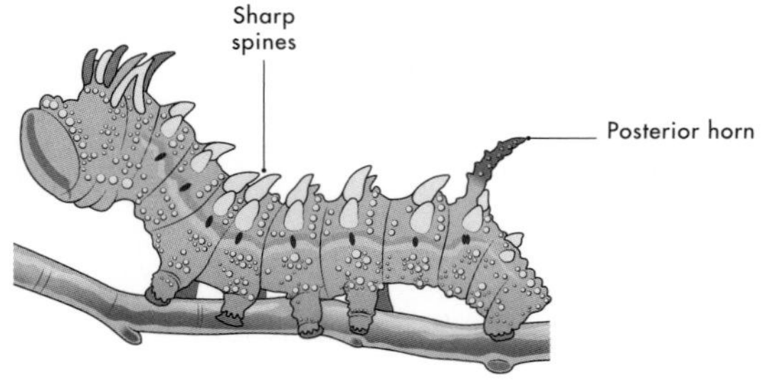

HELIOLONCHE PICTIPENNIS

Black-tipped Heliolonche

Fast-flying flower feeders

SCIENTIFIC NAME	*Heliolonche pictipennis* (Grote, 1875)
FAMILY	Noctuidae
NOTABLE FEATURES	Diurnal, black hind wings, sometimes with white stripe
WINGSPAN	⅝ – 11⁄16 in (16–17 mm)
SIMILAR SPECIES	Other *Heliolonche* species: *H. carolus, H. celeris, H. joaquinensis,* and, especially, *H. modicella*

Small and stout, like all noctuids, the Black-tipped Heliolonche is a fast-flying moth with red or light-yellow forewings, and black hind wings with a thin white border and sometimes a transverse white stripe. This diurnal species can be found in early spring in dry habitats of the southwestern United States, where its host plants grow—smooth desert dandelions (*Malacothrix glabrata*) and plumeseed (*Rafinesquia* spp.).

TUNNELING THROUGH FLOWERS

Females, which lay up to 40 eggs, insert them between florets above developing seeds. The larvae hatch five days later and feed on the flowers, eventually attacking the seeds; to fully develop, the cutworm-like larvae need to feed on two flower heads. Neonate larvae have an orange-brown head, while the mature fifth instar larva has a cream and chocolate-brown head, with a dark prothoracic shield that helps it tunnel through a flower. A flower-feeding caterpillar will spend the night inside the flower head as the inflorescence closes in the afternoon. It pupates in soil, and the pupa is smooth, with a cremaster modified into four thin spines, which may have a defensive purpose.

FLOWER MOTH RELATIVES

The most recent molecular study of the noctuid subfamily Heliothinae, to which *Heliolonche* belongs, suggests that the five *Heliolonche* species of similar desert-dwelling moths—all from the southwestern United States—are separate but closely related to the large genus of similar moths in the highly speciose genus *Schinia*, commonly called "flower moths" and mainly found in North America, but with a few representatives in Europe.

→ Black-tipped Heliolonche moths can be found zooming rapidly among the flowering host plants in early spring, when desert plants bloom.

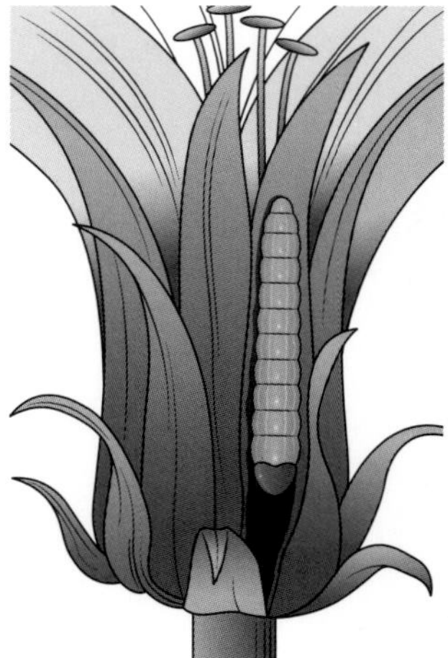

Tunneling larvae

Black-tipped Heliolonche moth larvae develop inside the flowers of their desert dandelion host plant, feeding on the flowers and finally the seeds.

PHYLLODESMA AMERICANA

American Lappet Moth

Leaf-shaped and cryptic

SCIENTIFIC NAME	*Phyllodesma americana* (Harris, 1841)
FAMILY	Lasiocampidae
NOTABLE FEATURES	Large, leaf-shaped moth
WINGSPAN	$1\frac{5}{16}$ – $1\frac{15}{16}$ in (29–49 mm)
SIMILAR SPECIES	Other *Phyllodesma* species, such as the Eurasian Aspen Lappet (*P. tremulifolium*) and also *Gastropacha* species

Typical of its genus *Phyllodesma*, named from the Greek *phyllon* ("leaf"), the American Lappet Moth is remarkably leaflike in its wavy shape and coloring. It is found throughout North America, from British Columbia and Yukon in Canada south to at least northern Florida and California. In the south, where two generations occur annually, its flight period begins as early as March and persists as late as September, while in the northern part of its distribution area, it produces one generation a year and flies between May and August.

SUCCESSFUL SPECIES

This species is highly polyphagous, which partly explains its success in colonizing a variety of latitudes, altitudes, and habitats—from deserts to northern and alpine tundra. The larvae feed on birches, oaks, cottonwood, willows, and various Rosaceae, among other plants. A 2012 study demonstrated that where areas of a forest have been stripped of trees, *P. americana* is among the fastest recovering moth species.

CRYPTIC ADULTS AND LARVAE

Other than its leaflike camouflage, the adult moth possesses few other defenses, being neither toxic nor having the ability to hear or produce sounds. The larvae, however, are both cryptic and covered in hairs that may serve a protective purpose. The name "lappet," which this moth shares with other Lasiocampidae species, derives from the hairy flaps on their prolegs. Lasiocampids are also known as eggars for the egg-shaped cocoons they produce, in which this species overwinter.

In addition to *P. americana*, there are six subspecies, several of which, such as *P. a. californica*, are named for the area in which they occur.

→ The American Lappet Moth is found as far north as Nova Scotia, British Columbia, and Yukon, and as high as the alpine tundra of Colorado. Its success in these extreme areas is largely due to the polyphagous nature of its larvae and their cold-hardiness.

HYLES LIVORNICA

Striped Hawk Moth

Spectacular pollinators

SCIENTIFIC NAME	*Hyles livornica* (Esper, 1780)
FAMILY	Sphingidae
NOTABLE FEATURES	Striped forewing, pink hind wings with olive-green borders
WINGSPAN	2⅜–3⅛ in (60–80 mm)
SIMILAR SPECIES	*Hyles lineata*

One of the world's most widespread moths, the Striped Hawk Moth can be found throughout Africa and Eurasia. The moths tend to breed in the southern part of their range and, in Africa and the Arabian Peninsula, can breed in the desert after rains. Migrants may also reach as far north as tundra, and have been recorded in Sweden and Novosibirsk in western Siberia. Occasional infestations of European grapevines have been recorded.

POLYPHAGOUS CATERPILLARS

Females can lay up to 500 eggs. Their larvae, which are green with black markings, defoliate entire desert plants and are highly mobile, frequently crawling some distance in search of a new host. They can develop on a variety of plants, such as sorrels and grapes, and, although not specialists on toxic species, unlike other *Hyles* species, they are adequately equipped to feed on toxic euphorbias that offer some protection against tachinid parasitoids that wipe out up to 80 percent of larvae.

IMPORTANT POLLINATORS

Striped Hawk Moths pollinate a variety of flowers—typically species that bloom at dusk and have a strong fragrance. In South Africa, they are essential pollinators of several orchids, which are uniquely adapted to transfer their pollinaria onto the moths when they insert their long proboscis into the flowers. The moths also visit less specialized flowers, such as cacti, thistles, morning glories, and evening primrose. As a result, this species has been used in chemical ecology studies to determine floral attractants—the visual and chemical cues the flowers employ to attract pollinators.

→ The Striped Hawk Moth feeds from flowers without landing, hovering like a helicopter above them.

Forewing pattern

The Striped Hawk Moth has pink hind wings that are dark at the base and along the margins; at rest, its cryptically colored forewings cover its hind wings.

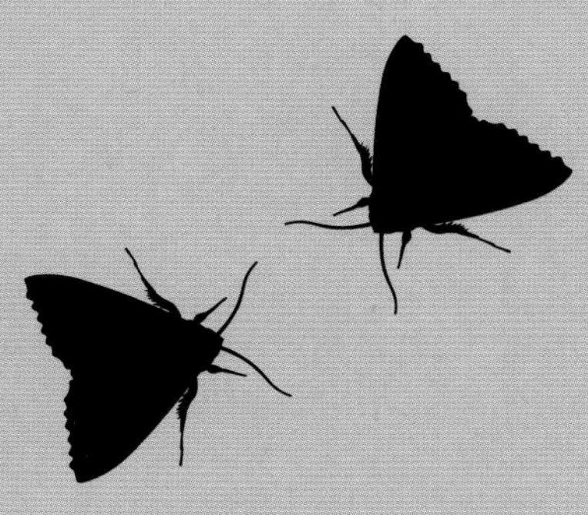

MOTHS OF TEMPERATE DECIDUOUS FORESTS

Spring abundance

The arrival of spring brings fresh life to the flora and fauna of temperate deciduous forests that occur in patches between southwest Canada and the southeastern USA, and across northwest Europe to China and Japan. Logged for centuries, these woodlands have a high capacity for regeneration and, in some parts of Europe, their area is even beginning to increase. They are also still found in the southern hemisphere, scattered in Chile and Argentina, southern Australia, and New Zealand's South Island. Moth development is timed to the forests' seasons: after overwintering in a dormant state as eggs, caterpillars, or pupae, the year's first generation arrives in full force in spring, and newborn larvae begin to feast on sprouting leaves.

TIMED TO EMERGE

In many parts of the northern hemisphere, oaks, birches, aspens, and maples are among the trees that dominate deciduous forests. The emergence of the many moth species associated with these and other broadleaf trees is biologically synchronized with the appearance of leaf buds and coincides with spring rains in the south,

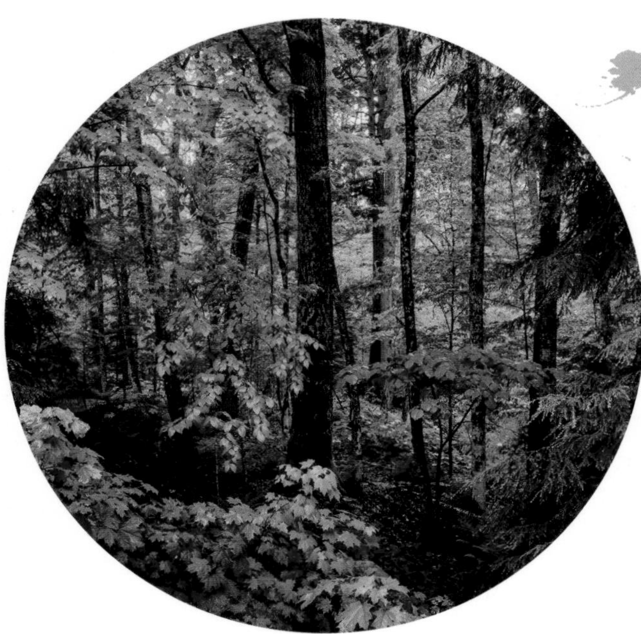

↑ A generalized biome map of temperate forest locations around the world.

← A deciduous forest in the spring, when newborn larvae emerge in force.

or with the snowmelt farther north and in the mountains. Overwintering eggs may be concealed on the bark of bare trees, and the caterpillars that hatch from them are among the first in line to feed on fresh growth, together with those that have spent the winter as larvae. Many moths that overwinter as pupae also eclose in the spring, quickly mate, and lay eggs to further swell the caterpillar population.

↑ A bilateral gynandromorph of the Kentish Glory moth (*Endromis versicolora*), Italy. This incredible specimen, half-male and half-female, is the result of successful fertilization and development of an aberrant egg with two nuclei.

Moths on oaks

The oak, which encompasses more than 500 species worldwide, is a stalwart of deciduous forests and vitally important to moths and other insects. *Quercus robur*, for example, the iconic English oak, found largely in Europe but also sporadically in the United States and China, supports more than 400 insect species, not to mention squirrels and wild boar that rely on its acorns for food.

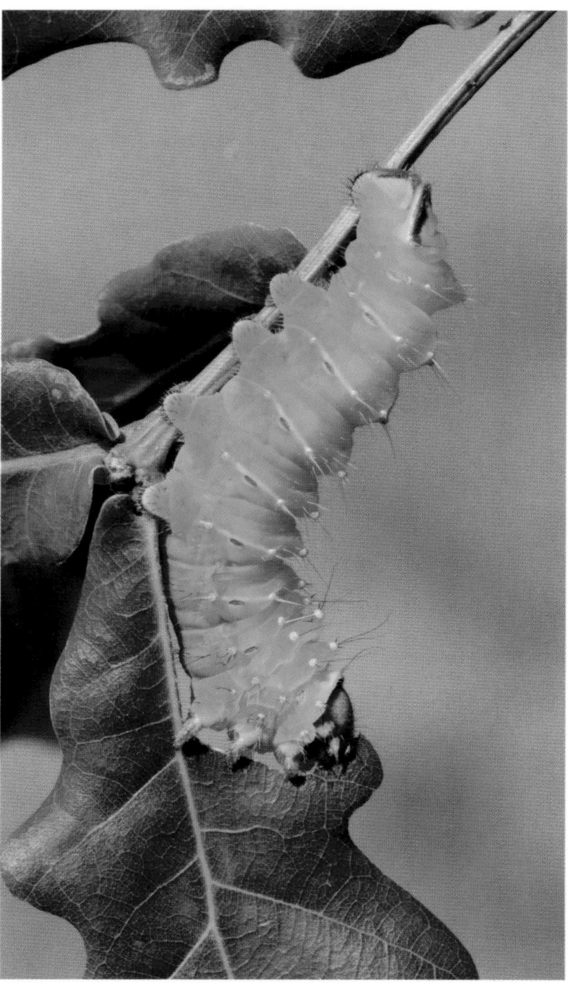

OAK-FEEDERS OF FLORIDA AND MISSOURI

More than 30 *Quercus* species grow in the southeastern United States, varying greatly in size and shape, many of them in the same habitat. As with other deciduous tree genera, their leaves become tougher and better defended by chemicals such as tannins as they mature, but in early spring they are a plentiful food source for caterpillars and poorly protected against their attack. Moth caterpillars may specialize on one species of oak, but more commonly feed on several species and sometimes on other trees, too.

Florida's butterflies and moths have attracted naturalists since the early twentieth century, when the state was a frontier of natural history knowledge. By the year 2000, the state's list of Lepidoptera had grown to almost 3,000 species, more than 200 of

↗ The Polyphemus Moth (*Antheraea polyphemus*), widespread in North America, has cryptic forewings that hide the defensive eyespot pattern of the hind wings. The adult moths do not feed.

← The Polyphemus Moth (*Antheraea polyphemus*) caterpillar feeding on oak. At pupation, it attaches its tough silvery silk cocoons to the branches, so that it and the pupa inside can survive the winter.

SACK-BEARER

In North America, the oak-feeding caterpillar of the Scalloped Sack-bearer (*Lacosoma chiridota*), a species from the obscure family Mimallonidae, creates an unusual netting structure on the skeletonized oak leaf when it is very young, and later prepares a silk-lined chamber from leaves, in which it lives and later pupates. The adults have a somewhat eccentric posture when at rest, with their wings pointing downward and abdomen upward, and most mimallonids resemble leaves with tiny eyespots, lines, and other features that assist in camouflage. A recent scientific paper suggests that Scalloped Sack-bearer moths mate by day, although the males appear to be attracted to female sex pheromones, a feature more commonly found in nocturnal species. Until recently, moths in this family were thought to be relatives of bombycoids, such as lappet and silk moths, which they resemble. However, a 2017 study confirmed that they belong somewhere between hook-tip moths and pyralids on the evolutionary tree.

which are known to feed on one or more different types of oak. Among them are the spectacular Io (*Automeris io*) and Polyphemus (*Antheraea polyphemus*) silk moths (Saturniidae). As young larvae, oak worms (saturniid genus *Anisota*), feed in large groups that frequently defoliate oak branches.

In the differing forests and climate of Missouri, about 1,000 miles (1,600 km) north of Florida, an abundance of moths also favor oaks. In 2018, a catalog of Missouri's oak-consuming caterpillars listed 107 moth species in 20 families. Among them are two species of Tortricidae: the Oak Leaf Roller (*Archips semiferanus*) and the pretty Oak Leaftier (*Acleris semipurpurana*), which can both badly damage oak foliage. The two species overwinter as eggs, with larvae that emerge in the spring and consume buds and young leaves. They then roll leaves together, securing them with silk, and hide inside them until pupation is complete and the adults emerge.

Other Missouri oak-feeders include members of the Notodontidae moth family, also known as prominent moths, whose cryptically colored, gray-brown adults look more similar than their distinctive caterpillars, which often have bold markings and colors, and sometimes tail-like or hornlike projections as in Mottled Prominent (*Macrurocampa marthesia*) and other prominent moths. Their "prominent" name derives from the tuft of long hairs at the edge of the forewing. When resting, they often curl their wings around their abdomen and lift their bodies off the ground to mimic twigs.

EUROPEAN OAK-FEEDERS

Numerous European moths use oaks as hosts. With the aim of recreating more biodiverse habitats, European Union initiatives encourage tree planting and natural forest regeneration. Where new secondary forest has regrown on abandoned pastures, oaks are among the first trees to re-emerge—and moths are quick to follow. In one such setting in Spain, a 2016 study described 22 species of moth caterpillars feeding on holm oaks (*Quercus ilex*).

← The Clouded August Thorn (*Ennomos quercaria*) is widespread in southern Europe; its inchworm larvae feed on oaks.

↓ Oak Processionary (*Thaumetopoea processionea*) caterpillars parade nose-to-tail to locate a new host plant after defoliating the previous one.

Among European oak-feeders are the Oak Processionary (*Thaumetopoea processionea*), named for the head-to-tail processions of its larvae as they move between plants, often defoliating oak trees. The beautiful, cryptically colored Clouded August Thorn (*Ennomos quercaria*), an oak specialist as its species name suggests and found mostly in southern Europe, has recently surfaced in southern England, possibly moving north as a result of global warming.

European species with oak-feeding caterpillars include the Spiny Hook-tip (*Drepana uncinula*) and the Oak Hook-tip (*D. binaria*). The Drepanidae moth family encompasses more than 600 mostly forest-dwelling species, many of which superficially resemble geometrids as adults, as they have similar flatly positioned, broad wings and slender bodies, but can sometimes be recognized by their extended and rounded hooklike forewings. Their bulky caterpillars, however, have all their prolegs, unlike slender geometrid inchworms, which have a reduced number of abdominal prolegs. Drepanid rear prolegs are frequently modified into projections and lifted off the substrate.

SUBTLY COLORED "UNDERWINGS"

Of the 250 species of the moth genus *Catocala*, most live in temperate forests and overwinter as eggs. Their cryptically colored, flattened, twiglike caterpillars feed on deciduous trees such as poplar, oak, willow, and walnut, while the fast-flying adults sip sugar-rich tree sap, which seeps out where other insects have penetrated the bark. The moths are known as "underwings" for their beautiful yellow, red, or blue hind wings, tucked under the drab coloring of the forewings, which are a perfect camouflage when they land on the bark of older trees. The larvae of many underwing moth species are oak-feeders. The color and pattern of caterpillars of the North American Ilia Underwing (*Catocala ilia*) resemble the lichen that grow on their oak host plant. The small scrub oak (*Quercus ilicifolia*) of the eastern United States and southeastern Canada, is one of several hosts for larvae of the Andromache Underwing (*C. andromache*), a large and beautiful moth with yellow hind wings hidden. European oak-feeders include the crimson underwings (*C. promissa* and *C. sponsa*).

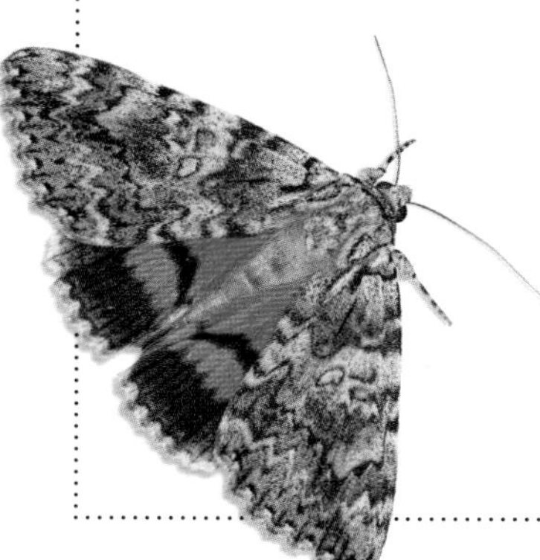

↑ *Catocala* moth caterpillars are masters of camouflage, resembling bark and, sometimes, lichen.

← A *Catacola promissa* is one of several similar underwing moths found in the Holarctic region that encompasses Eurasia and North America. Its larvae feed on oaks.

Moths on maples

Like oaks, maples of the genus *Acer*, of which there are more than 130 species, constitute a large and diverse food resource for moths. The sugar maple (*Acer saccharum*), from which maple syrup is gathered in spring, is one of the best known. Sugar maples grow across a large area of northeast America, extending south to northern Virginia in the east across to Missouri in the Midwest.

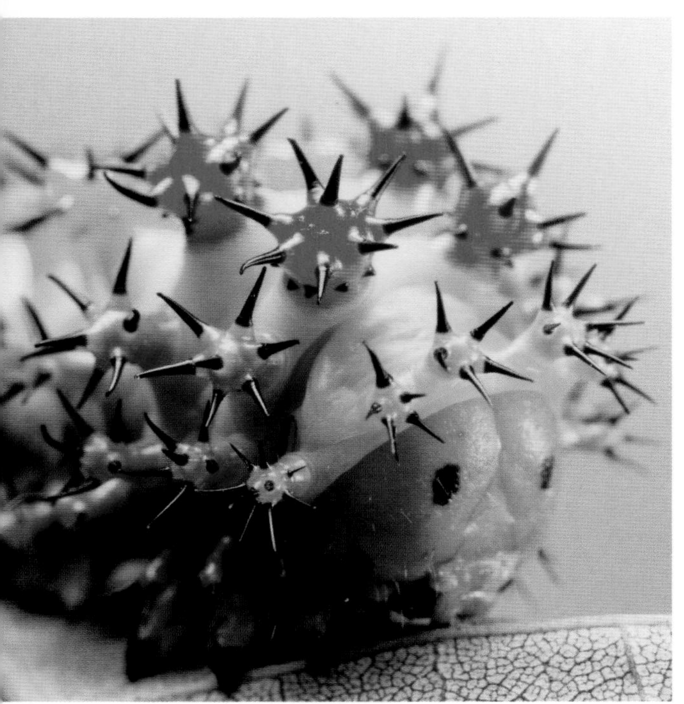

↑ The Cecropia Moth caterpillar (*Hyalophora cecropia*), defended against predation with spines, prefers maples among deciduous trees, though has been recorded feeding on plants from over 20 different families.

SPECTACULAR MAPLE FEEDER

Among the moth caterpillars that feed on sugar maples and other Acer species are larvae of the Rosy Maple Moth (*Dryocampa rubicunda*), a spectacular pink-and-yellow, highly variable silk moth. The female lays eggs in batches of 20 to 40 on the underside of maple leaves and, in high concentrations, its larvae occasionally defoliate red, silver, and sugar maple trees. The Cecropia Moth (*Hyalophora cecropia*), another giant silk moth, also favors maples, but can develop on a variety of other plants, including many temperate forest tree species.

WOOLLY BEARS AND INCHWORMS

Many moth caterpillars that can feed on multiple trees include maple leaves in their diet. Among them is the hardy Banded Woolly Bear (*Pyrrharctia isabella*), which overwinters in freezing temperatures, protected by a substance that safeguards its tissues. Other maple-feeders include the Yellow Woolly Bear (*Spilosoma virginica*), the Pale Tiger Moth (*Halysidota tessellaris*) caterpillar, and polyphagous inchworms, such as the twig-mimicking larvae of the Crocus Geometer (*Xanthotype sospeta*) and the Pale Beauty (*Campaea perlata*). All five species avoid the first line of leaf defenses (the tough leaf cuticle and epidermis) by making a hole in the surface layers and feeding on the inner leaf.

Researchers studying the feeding behavior of these five moths on two Acer species—the sugar maple and

the boxelder maple—discovered that, as the caterpillars matured, their ability to feed on the two maples clearly differed. The two generalist moths—*P. isabella* and *X. sospeta*—that also feed on herbaceous plants readily ate leaves of the boxelder maple, while the other three species that prefer trees, fared better on sugar maples. Analysis of the leaves of the two maples for ten different groups of protective secondary plant compounds, such as alkaloids, benzenoids, and coumarins, revealed that most were present in the boxelder maple, but the sugar maple had only two. From this, it was evident that the two generalists that consume a variety of plants found it much easier than the tree-feeders to overcome the boxelder's many chemical defenses.

SLENDER MAPLE-SLAYERS

Geometrid caterpillars often remain unnoticed by impersonating twigs but on occasions become all too apparent. Outbreaks of the North American Elm Spanworm (*Ennomos subsignaria*) can defoliate scores of sycamore maple trees, as occurred in eastern Canada between 2002 and 2006. Such outbreaks (sudden increases of population that last for several generations) are a normal part of insect population fluctuations and are ultimately controlled by correspondingly large numbers of their predators and parasitoids, but nevertheless can have a devastating effect. Researchers have noted that, during these outbreaks, certain forms of *E. subsignaria* caterpillars, such as melanics (dark-colored larvae) or ones with rusty-colored heads, become more prevalent. Trees react to a major attack by producing more protective tannins in their leaves, which probably influences the caterpillars' coloring. A 2009 study of timing and distribution of the Elm Spanworms showed that they survived best on slightly older foliage, preferably in the crown of the tree.

↖ ↗ Rosy Maple Moth caterpillars (*Dryocampa rubicunda*) (top left) are common across the eastern United States and southern Canada in temperate deciduous forests and suburban areas and even occasionally in cities, where they feed on maples and, infrequently, on oaks. The adults (top right) do not feed and are highly variable in their coloration.

Cryptic geometrids

Inchworms (Geometridae) are among the most abundant moths in temperate deciduous forests, and are heavily predated by many creatures, including birds, wasps, and spiders. The 23,000 species worldwide—more than 1,400 of them in North America—have evolved numerous survival strategies, but with some general characteristics. With a few exceptions, inchworms and their adult moths are protected only by their cryptic coloration. Most of these moths rest with their wings held open and flat, displaying highly patterned but plain-colored forewings and, usually, similarly colored hind wings, which blend perfectly with bark or leaf litter, frequently mimicking a lichen or a dry leaf.

HIDDEN IN PLAIN SIGHT

Inchworms are well camouflaged on their host plants, often disguising themselves as twigs when feeding on a number of deciduous trees. Among the best stick mimics are Purplish-brown Loopers (*Eutrapela clemataria*) that overwinter as pupae and whose adults fly early in spring. Caterpillars of the Red-fringed Emerald (*Nemoria bistriaria*) may resemble the fringe of a dry oak leaf, while the adult moths are colored green or ocher, to match both dry and living leaves. *Ceratonyx satanaria*, which develops on sweet gum (*Liquidambar styraciflua*), got its species name for the two hornlike projections behind its head that look like a forked branch and likely have additional sensory functions. Like many other inchworms, it also has fake leaf scars along its long, slender body in places where the segments connect, which add to the twig effect. *Ceratonyx satanaria* adults have slender wings and hold their abdomen pointed at right angles to the thorax, which is more typical of crambid moths. Adult Oak Beauty Moths (*Phaeoura quernaria*) resemble the polymorphic Peppered Moth (*Biston betularia*), but can be even more variable, ranging from completely black to almost white to match a variety of complex

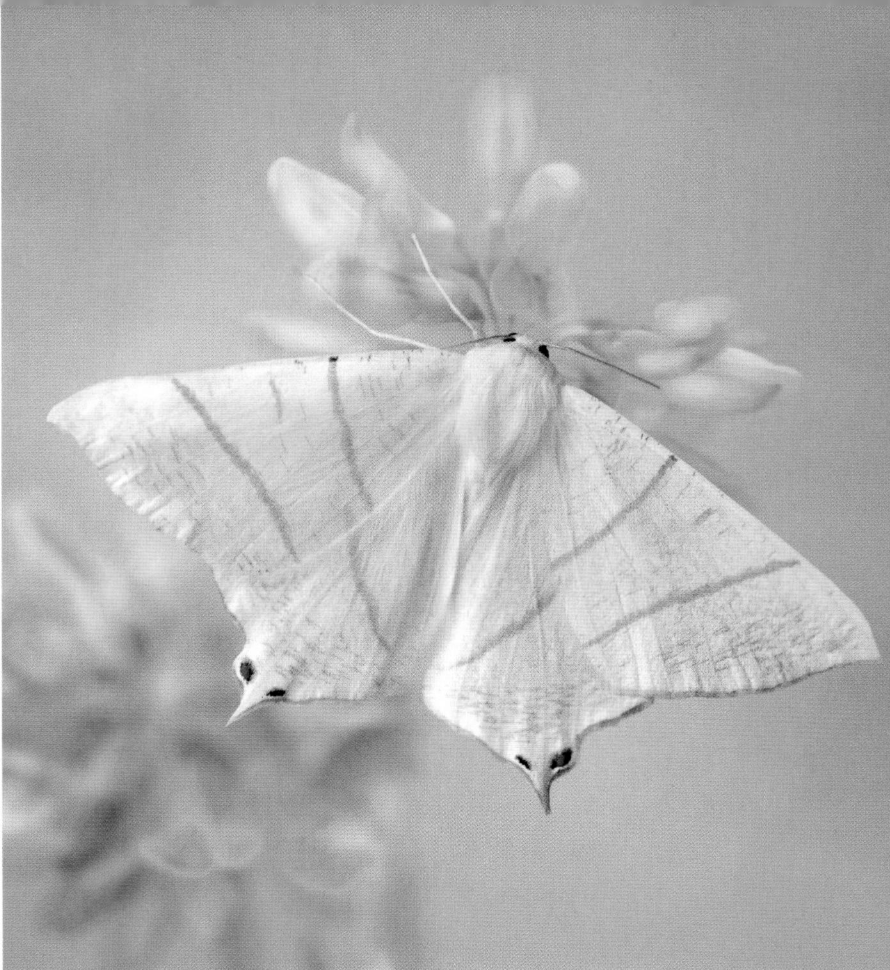

→ The markings of the Swallow-tailed Moth (*Ourapteryx sambucaria*) are an example of false-head patterning, with tails and eyespots to deflect the attention of predators from its head to its hind wings.

← Most inchworms are cryptic and, like the Oak Beauty Moth (*Phaeoura quernaria*) larva here, mimic a twig of their host plant.

types of tree bark on which they rest. As in most geometrids, males and females are hard to tell apart, but in some, such as the Mottled Umber (*Erannis defoliaria*), females are wingless.

VARIED DIETS

Despite its name, the leaf-mimicking Oak Besma (*Besma quercivoraria*), widespread from Canada to Mexico, has sticklike, thin, gray larvae that feed on oaks and a variety of other trees. The Pale Beauty Moth (*Campaea perlata*) caterpillars have even more varied diets and have been recorded on 65 different species of trees and shrubs in deciduous forests from Canada to Arizona. Two species that are widespread in Europe—the Brindled Pug (*Eupithecia abbreviata*) and March Moth (*Alsophila aescularia*)—together with the North American Linden Looper (*Erannis tiliaria*), can also develop on a variety of trees, and their forewing colors and patterns blend with a bark background, although *A. aescularia* and *E. tiliaria* females are wingless.

In Europe, one of the loveliest geometrids, frequently found in temperate forests, is the white Swallow-tailed Moth (*Ourapteryx sambucaria*), which has a wingspan of up to 2⅓ in (60 mm). Its slender, sticklike caterpillars feed on various trees and shrubs, such as hawthorn or honeysuckle. In Britain and farther south in Europe, the Magpie Moth (*Abraxas grossulariata*) has a distinctive white-yellow-black speckled pattern, and similarly colored caterpillars and pupae. The larvae feed on the blackthorn bush (*Prunus spinosa*), as well as several other trees and shrubs, such as willows and gooseberries. The moth contains the bitter compound sarmentosin, which protects it from predators.

Caterpillar silk

Many moth caterpillars have the capacity to create silk via their silk glands, laying it down with the help of a modified mouthpart called a spinneret. Caterpillars use this silk in many ways, for instance, to bind leaves, construct cocoons, and to propel themselves through the air. Tree-feeding caterpillars, most numerous in spring, frequently swing down from branches: some drop down to avoid predation, while others, such as larvae of the Laurelcherry Smoky Moth (*Neoprocris floridana*), use silk to swing from a defoliated host tree to another less damaged one in search of food.

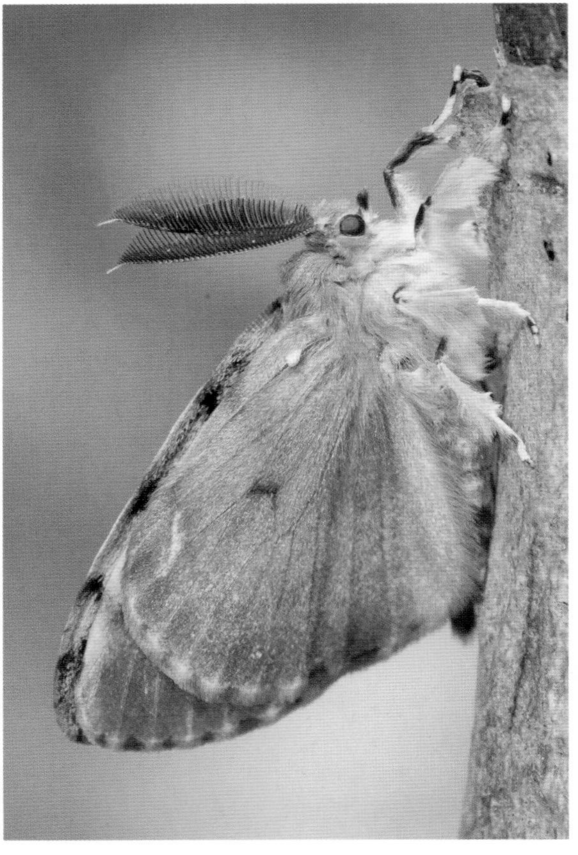

GYPSY MOTHS

Lymantria, a vast genus of over 200 species, includes the infamous Gypsy Moth (*L. dispar*), whose newborn larvae are largely responsible for its dispersal and achieve this by "ballooning"—flying through the air on silk threads (see also page 27). The hairy mature larvae attack a wide variety of plants and, during outbreaks, have defoliated large swathes of temperate forests throughout the world. The Gypsy Moth's white-patterned females are larger than the brown males, hence the name "dispar," meaning "different" in Latin. In European and North American populations,

← Gypsy Moth (*Lymantria dispar*) males look for females, following a trail of pheromones, which they pick up with sensilla on their serrate (featherlike) antennae. Females of the European subspecies are largely immobile, while those of the Asian subspecies can fly.

↗ Tussock moth caterpillars in the genus *Orgyia* are protected with tufts of hairlike setae, which can produce an allergic reaction if touched. Pupating caterpillars also weave the hairs into their cocoons.

the females are mostly sedentary, but sporadically fan their wings and occasionally walk. However, in Asian populations of the same species, females are more active, dispersing mostly short distances (within a mile) when laying eggs. Occasionally they also migrate to cooler mountain forests before they mate.

WHEN IT RAINS CATERPILLARS

The larvae of White-marked Tussock Moths (*Orgyia leucostigma*), like those of Gypsy Moths with which they share the same Lymantiinae subfamily, are highly polyphagous, feeding on hundreds of different plants. They and their close relative, the Fir Tussock Moth (*O. detrita*) that specializes on live oaks and bold cypress, are numerous during springtime in the southeastern United States. Their larvae drop from trees when mature and ready to pupate, sometimes in large numbers, and spin silk cocoons everywhere, including on man-made structures. The cocoons are constructed not only from silk, but also from caterpillar hairs that have irritating properties. The yellow tufts on the caterpillar's back also contain urticant hairs, sometimes reinforced by poison glands, which can cause an allergic skin reaction and even welts if brushed against soft skin. The brown cryptic males of these species search for flightless females using their pheromone-sensitive, comb-shaped antennae. The females have greatly reduced, rudimentary wings and resemble little cotton balls. They stay on their cocoon after emerging, then lay around 300 eggs on its silky surface; the young caterpillars disperse the following year.

LIVING IN TENTS AND NESTS

Making nests is generally associated with vertebrates and, among arthropods, with ants, wasps, or social spiders. Lepidoptera might seem unlikely nest-builders, but there are great benefits to living communally, so this strategy has evolved in the larvae of many butterflies and moth species. In Lepidoptera, nests usually result from a single large batch of eggs laid by one female, though in some butterflies, females have also been observed combining their egg-laying efforts, so that more caterpillars emerge together and are able to build an even larger nest.

TENT CATERPILLARS

In temperate forests across the world, lappet moths or eggars, Lasiocampidae, such as the European Lackey Moth (*Malacosoma neustria*) and the American Forest Tent Caterpillar and Eastern Tent Caterpillar (*M. disstria* and *M. americana*), overwinter as egg batches that circle tree branches. Once hatched in early spring, young *Malacosoma* caterpillars combine their ability to overcome plant defenses and continue to live together for their mutual benefit in tents that they spin around host-plant branches.

TENTS ON APPLES AND ULTRASONIC CLICKS

The speckled white Bird-cherry Ermine (*Yponomeuta evonymella*), a micromoth in the family Yponomeutidae, has a very similar relative, the Apple Ermine (*Y. malinellus*), infamous for its caterpillars' ability to devastate apple orchards. The larvae of both species cloak their host trees with communal webs and feed under them, protected from predators. While small, their collective action results in vast amounts of silk that can envelop branches and sometimes entire trees. These ermine moths can reach huge numbers in Europe's forest edges and parks, and may completely defoliate trees and occasionally drop on passersby. In the sixteenth century, Austrian monks used strands of caterpillar silk glued together to create a fine

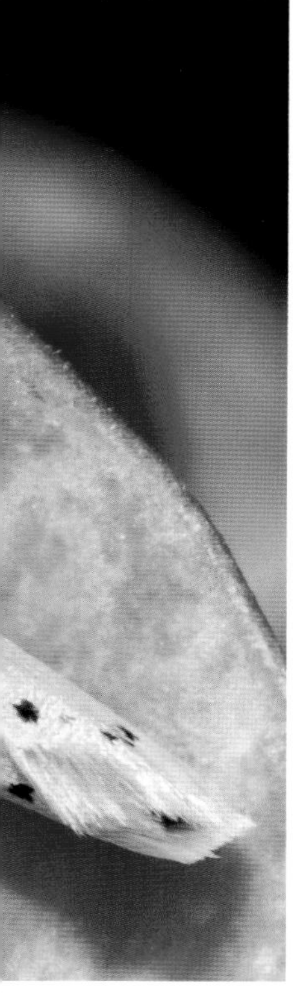

net canvas on which they painted religious miniatures. One of the very few examples that still exist is an image of the Virgin Mary in Chester Cathedral, England, painted on the silk of *Y. evonymella*.

While it is their destructive behavior that attracts the attention of researchers, tiny ermine moths are of further scientific interest. Research on the Apple Ermine has shown, for instance, that the proximity of a host plant stimulates pheromone production in females. A 2019 study on the Bird-cherry Ermine and the Spindle Ermine (*Y. cagnagella*) also described an interesting acoustic behavior —it appears that these moths are deaf but constantly produce ultrasounds using clicks of their wings to repel bats, mimicking the ultrasounds of toxic tiger moths.

↖ ↑ The genus *Yponomeuta* includes over 100 similar moth species. Their caterpillars (top) are known to envelop trees and shrubs (the common spindle here) with abundant silk. The look-alike moths are most reliably distinguished by their host-plant preference.

← Lackey Moth caterpillars (*Malacosoma neustria*) that emerge from clusters of overwintering eggs create "nests"—silken nets under which they shelter together in large groups and feed.

Silk moths and relatives

Bombycoidea is a superfamily of larger moths that includes a number of spectacular species: silk moths, hawk moths, lappet moths, and others, many of which are found in deciduous forests. Despite their size, they and their caterpillars are rarely visible in the forest by day. Being large and slow fliers, but defenseless against birds and mammals, they remain dormant and fairly immobile during the day, so, of necessity, they are among the greatest masters of camouflage and other cunning survival strategies.

SPECTACULAR "LEAVES" WITH EYESPOTS

Silk moths may rely on leaf- or bark-mimicking forewings, but sometimes have a backup strategy—brightly colored hind wings that flash at a predator when the moth is disturbed. In the temperate deciduous forest of northern Japan, many silk moth caterpillars munch on young leaves of mizunara—the Japanese oak species (*Quercus crispula*). Among them are the Jonasi Silk Moth (*Caligula jonasi*) and the tailed Artemis Moth (*Actias artemis*), whose North American relative, the Luna Moth (*A. luna*), feeds mainly on sweet gum in the southeast of the United States. A 2018 study of moths and captive bats that used high-speed videography showed that the longer the tails in silk moths the better they are at deflecting bat attacks. Luna Moths have a uniform light green coloring that blends perfectly with their surroundings when they are at rest during the day, but they also have translucent eyespots that are much more noticeable to predators when viewed against the light.

The Imperial Moth (*Eacles imperialis*) is also cryptic at every life stage, and its caterpillars can be either brown or green, depending on their genetics.

→ Caterpillars of the Imperial Moth (*Eacles imperialis*) exhibit polymorphism—they have genetically determined color forms, each of which would provide a better camouflage and thus predator avoidance under different circumstances (lighting, background, and so on).

← The North American Luna Moth (*Actias luna*) can be recognized by its distinctive hind wing shape and tail length.

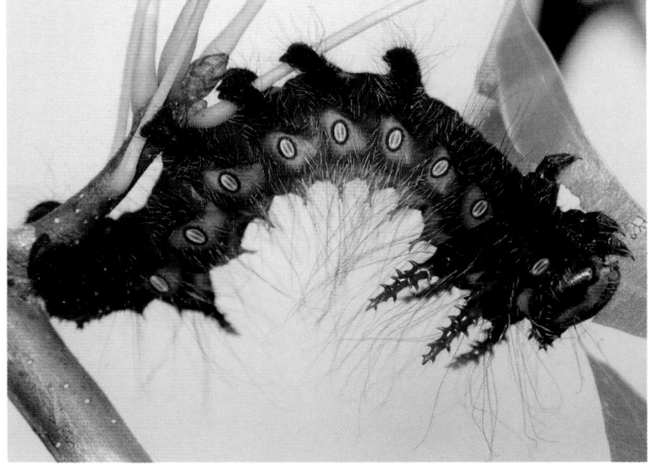

Once they finish feeding on their tree host plants, including oaks and maples in the southeastern United States and the pitch pine in New England, Imperial Moth caterpillars crawl down the trunk and bury themselves in the ground, where they build a chamber and pupate without a cocoon. As adults, they resemble large, dry leaves with intricate, frequently variable patterns. A 2010 study of the New England population indicates that their significant decline in the last half-century may be partly due to the loss of their host-plant trees.

KENTISH GLORY

The European forest moth, the Kentish Glory (*Endromis versicolora*) heralds the spring as males begin to fly early in March and, attracted by the females' pheromones, search them out by day and night. They form just one generation a year. Males are not only smaller, but also darker than females, with ocher hind wings. The egg-laden, more sedentary females are well camouflaged against the bark of their birch host trees, blending in with scars and lichen, while the caterpillars are invisible among the foliage. The Kentish Glory is the only European member of its family Endromidae, which includes several additional genera from Southeast Asia.

HAWK MOTH LEAF MIMICS

The characteristics of many hawk moths (sphinx moths) found in temperate deciduous forests are much more like those of silk moths than the hawk moths of grasslands, desert, and rainforest, which are often migratory, airplane-shaped, and hummingbird-like. Unlike those of other hawk moths, the proboscis of forest species is rudimentary, and these moths do not feed as adults. Some hawk moths, whose caterpillars specialize on temperate deciduous trees, are also poor fliers. Like silk moths, their larvae pupate underground in chambers, their females hatch with eggs ready to lay, and many hawk moths have eyespots or other bright coloration on their hind wings. These they expose only

↖ The Kentish Glory (*Endromis versicolora*) is the only European representative of a tiny family of bombycoid moths and the only member of its genus. This species is a sign of spring in the temperate broadleaf forests of Europe, where its larvae develop on birch.

↑ The Lime Hawk Moth (*Mimas tiliae*) is named after its host plants (genus *Tilia*) known as basswood in North America and linden in Europe. The "Lime" in its common name comes from the British use of "lime tree" for *Tilia*—there is no connection with citrus.

when threatened, otherwise hiding them under cryptically colored forewings that are shaped and colored to resemble an elongated dry leaf—also a feature of many silk moths.

Cryptic, leaf-shaped caterpillars of the Poplar Hawk Moth (*Laothoe populi*) and the Aspen Hawk Moth (*L. amurensis*), are found on poplar and aspen in European forests. Similar Oak Hawk Moth (*Marumba quercus*) and Lime Hawk Moth (*Mimas tiliae*) larvae feed on oak and basswood, respectively. While the Poplar Hawk Moth adult lacks eyespots, it has specks of red color on its hind wings, which are normally hidden. It also employs another cryptic strategy: its broad hind wings under its narrow forewings protrude above them, so that the outline no longer resembles that of a moth.

THE CATALPA SPHINX

Uncharacteristically for sphinx moths, caterpillars of the Catalpa Sphinx (*Ceratomia catalpae*) feed in groups on a single host-plant genus—*Catalpa*, pictured, also known as cigartrees or Indian-bean-trees for the shape of their seedpods. Prized as ornamentals for their beautiful flowers, they are medium-sized trees native to deciduous forests of the southeastern United States. To defend itself against Catalpa Sphinx attacks, the southern catalpa (*C. bignonioides*) produces sweet secretions that summon nearby ants to defend it. The plant also produces the iridoid glycosides catalpol and catalposide—defense chemicals, which Catalpa Sphinx caterpillars sequester for their own protection.

↓ The Catalpa Sphinx (*Ceratomia catalpae*) caterpillar is uniquely adapted to feeding on this chemically defended plant.

Armed, dangerous, and eccentric

Certain temperate deciduous forest moths and, more frequently, their caterpillars have an eccentric appearance and fascinating behavioral adaptations—ranging from bizarre, defensive postures and unusual shelters, to sound production and the ability to spray a toxic chemical or even slay wasps.

HIDDEN WEAPONS

While aspects of every moth species are unique in some way, certain temperate deciduous forest species have an especially odd appearance. The caterpillar of the Southern Flannel Moth (*Megalopyge opercularis*), for instance, resembles a furry, fairy-tale mammal, rather than an insect, although its cute appearance and fuzzy hairs belie toxic spines hidden among softer setae that cause a painful, and sometimes dangerous, sting. The mature caterpillar uses the soft setae to make a durable cocoon, topped with a pocket full of soft hairs, which may be an additional defense, as a bird pecking at it and receiving a mouthful of hairs might abandon any further attack on the cocoon. The cocoons are so tough that only a sharp knife can open them, and they remain on trees long after the moth has eclosed.

Larvae of the family Limacodidae may be short, flat, rounded, or even branched. The common name for them is slug moth caterpillars, as they have no hooks on their

LOBSTER MOTH

The caterpillar of the lobster moth (*Stauropus fagi*), which is widespread in Europe, got its name for looking rather like a bizarre crustacean that has climbed a tree. These caterpillars can feed on a variety of forest trees, from maples and oaks to willows and basswood, and the resulting hairy adult moths (that resemble bombycoids rather than notodontids) can have a wingspan of up to 2¾ in (70 mm).

prolegs, but instead use suction cups and liquid secretion to adhere to the top of a leaf, moving very much like slugs. In temperate forests of the southeastern United States, the many slug moths include the Saddleback Caterpillar moth (*Acharia stimulea*), the Hag Moth (*Phobetron pithecium*), and the Spiny Oak-slug Moth (*Euclea delphinii*). All of the caterpillars have stinging setae, backed up by venom of different levels of potency.

WASP SLAYERS

The Wasp Parasitizer Moth (*Chalcoela pegasalis*) and its relative, the Sooty-winged Chalcoela (*C. iphitalis*) lay eggs on paper wasp nests of the genus *Polistes*, and its caterpillars feed on immature wasps. These crambids have iridescent spots on the margins of their hind wings that resemble the eyes of a jumping spider—perhaps to deflect attacks. Tiny moth cocoons develop inside cells of the wasp nest, next to the remnants of their prey. A small wasp nest can yield several dozen moths.

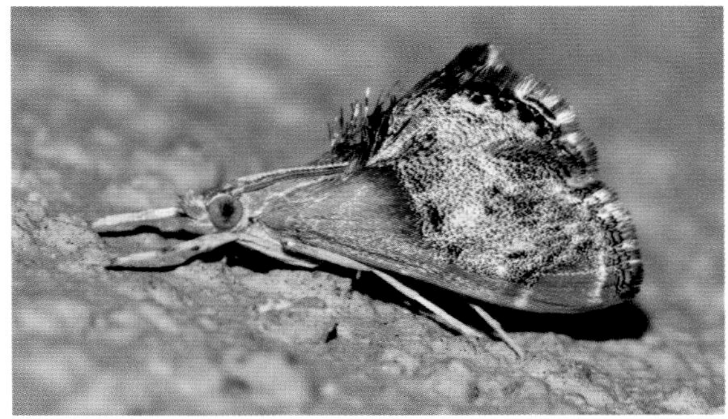

↑ The Sooty-winged Chalcoela (*Chalcoela iphitalis*) as larvae are predators on the immature stages of the *Polystes* paper wasp; a single nest can produce numerous moths.

← Well-armed with poisons and spines, caterpillars of two limacodid slug moths: the Smaller Parasa (*Parasa chloris*) (left) and the Saddleback Moth (*Acharia stimulea*).

AUTOMERIS IO

Io Moth

Stunning eyespots

SCIENTIFIC NAME	*Automeris io* (Fabricius, 1775)
FAMILY	Saturniidae
NOTABLE FEATURES	Large eyespots on hind wings, female's forewings darker than male's
WINGSPAN	2–4 in (50–100 mm)
SIMILAR SPECIES	Other *Automeris* species, such as *A. louisiana* (coastline of Gulf of Mexico)

Leaflike at rest—with yellow forewings if male and brown if female—the Io Moth hides the beautiful eyespots on its hind wings. If startled, it flips its forewings upward, suddenly exposing "eyes" that can rival those of a raptor in size and brightness, scaring off small bird predators. This defensive behavior is a bluff, as the moth has no chemical protection and cannot defend itself with rapid flight. These moths do not feed as adults as their proboscis is greatly reduced. They live for only one to two weeks.

BREEDING HABITS

The species is distributed from Costa Rica to Canada and, depending on location, can breed continuously or go into winter diapause. Males use their serrate antennae to locate females, following the pheromone trail they produce. The females, which mostly take flight only after mating, can lay around 300 eggs in batches of 20 to 30, so must visit at least 10 to 15 sites.

SPIKY, STINGING CATERPILLARS

The caterpillars are initially brown and feed in groups on a variety of plants, from oak and birch to willow and cherry. Later they become green with a candy stripe of white and red, and feed alone. Rather than deriving their defenses from plant chemicals, they produce their own toxins via glands that are connected to their numerous syringe-like spines, which can deliver a painful sting. When mature, the larvae spin a thin, strong cocoon, pulling together leaves for added protection, so the cryptic pupa can overwinter in harsh northern climates or survive prolonged droughts.

→ The male (top) is smaller and has lighter forewings than the female. The hind wing eyespots are composed of wide shingle-like scales in light and dark colors that give them an iridescent appearance, with similar pure white scales in its UV-reflecting center. The black ring of the eyespot and the rest of the wing are covered with longer, narrower scales.

Male

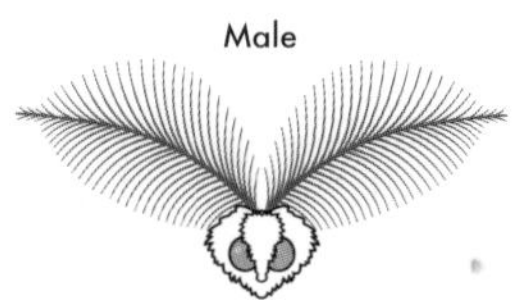

Female

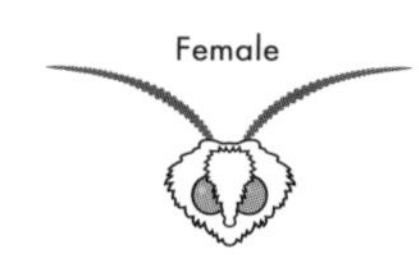

Sensitive antennae

The sexual dimorphism (difference between males and females) extends beyond wing coloration in the Io Moth and many other silk moths. The male's antennae, for instance, are wider with more sensilla, which enables him to locate a female following the vague trail of pheromones she releases.

PAONIAS EXCAECATA

Blinded Sphinx

Cryptic forest silk moth

SCIENTIFIC NAME	*Paonias excaecata* (Smith, 1797)
FAMILY	Sphingidae
NOTABLE FEATURES	Eyespots on hind wings
WINGSPAN	2³⁄₁₆ – 3¾ in (55–95 mm)
SIMILAR SPECIES	Holarctic *Paonias* and *Smerinthus* spp., such as the One-eyed Sphinx (*S. cerisyi*), which has an overlapping North American range

Named for the eyespot on each hind wing that lack a central dark "pupil," the Blinded Sphinx is found throughout North America, from Nova Scotia to British Columbia in Canada south to at least the northern parts of Mexico.

DRY LEAF MIMICS WITH "EYES"

Some characteristics of this and other hawk moths in the subfamily Smerinthinae resemble those of silk moths: they are cryptic, simulating a dry leaf in color and forewing shape, they do not feed as adults, they are forest dwellers with tree-feeding caterpillars, and do not fly especially fast. Like many silk moths, rather than fleeing, they also rely on the iridescent eyespots of their brightly colored hind wings for defense, exposing them to scare off predators if disturbed.

CRYPTIC CATERPILLARS

Paonias excaecata caterpillars, like other hawk-moth larvae, termed "hornworms" for the spike on their last abdominal segment, feed on trees and shrubs, including willows, birch, and cherry. The larvae are mostly green with thin stripes resembling leaf veins, but in late instars may have bright red spots, similar to those that some trees, especially cherry *Prunus serotina*, develop on the underside of their leaves, which is where the larvae feed, clinging to a protruding central vein. Rather than spinning a cocoon, they bury themselves underground to pupate, sometimes overwintering.

ONE OF MANY

The genus *Paonias*, which includes only four species, is a part of a larger complex of species with similar characteristics, collectively called Smerinthini, one of the three tribes within Smerinthinae.

→ The Blinded Sphinx in its defensive posture that exposes hind wing eyespots to ward off predators.

Powerful prolegs

Cryptic caterpillars of *Paonias* and other smerinthine hawk moths rest and feed on the underside of a host-plant leaf, holding onto the central leaf vein with powerful rear prolegs.

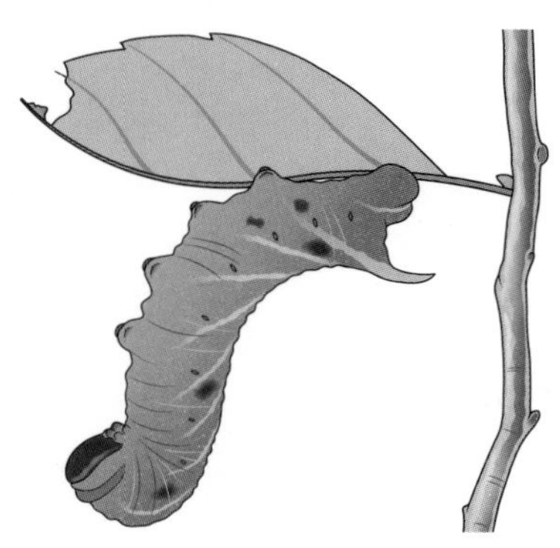

ENNOMOS MAGNARIA

Maple Spanworm Moth

Clever leaf and stick mimic

SCIENTIFIC NAME	*Ennomos magnaria* (Guenée, 1858)
FAMILY	Geometridae
NOTABLE FEATURES	Toothed wing margin, mimicking a leaf
WINGSPAN	1 11⁄16 – 2 3⁄8 in (43–60 mm)
SIMILAR SPECIES	Canary-shouldered Thorn Moth (*E. alniaria*) (Southwest Canada); several European *Ennomos* spp., such as the Large Thorn (*E. autumnaria*)

Shaped and colored like golden fall leaves, with variable dark spots and thin lines on their forewings, Maple Spanworm Moths are among the larger geometrids. Flying from July to November in one generation a year, they are found coast to coast in North America, from southern Canada to northern California in the west and to the Florida panhandle in the east. The snow-white Elm Spanworm Moth (*Ennomos subsignaria*) occurs within the same range in North America.

PERFECT STICK MIMIC

Overwintering as eggs laid on their host plant, the larvae hatch and feed on the leaves of alder, ash, basswood, birch, elm, hickory, maple, oak, or poplar from May to August, depending on location, and pupate by pulling leaves together to form a loose cocoon. The young larvae are green, but when mature they are among the best stick mimics of all caterpillars, with a base color varying from dark green to gray, and always with ridges, resembling leaf scars, where some segments connect.

FOOD FOR BIRDS

Despite their camouflage, these and other inchworms are frequently an important part of the diet of many birds and also predatory wasps that eat them or paralyze them to feed to their young—an invaluable (if unwilling) contribution to their ecological communities. Enough survive, however, ensuring that *E. magnaria* numbers are not under any threat.

→ The Maple Spanworm Moth freshly emerged from its pupa in Maine.

CATOCALA FRAXINI

Blue Underwing

A burst of blue

SCIENTIFIC NAME	*Catocala fraxini* (Linnaeus, 1758)
FAMILY	Erebidae
NOTABLE FEATURES	Blue band on hind wings
WINGSPAN	3½ – 4⅜ in (90–112 mm)
SIMILAR SPECIES	The White Underwing (*C. relicta*—southern Canada to Arizona and Missouri), also with blue underwing colors

At rest, with its wings held flat, *Catocala fraxini* conceals the striking band of blue on its hind wings that gives the species its common name. It flashes it when disturbed, while its forewings camouflage the moth when it rests at low levels on tree bark.

WIDE-RANGING SPECIES

All underwings are highly alert moths and have an acute sense of hearing. The species' range extends from Japan through Russia, and from Central Asia to Turkey and across Europe, and some populations have lighter, almost white forewings. The moth is often common, flourishing where aspens and poplars (its larval host plants) grow, and it prefers warmer climates, so is relatively rare in northern Europe. It ceased to breed in the UK for a period from the 1960s, as a result of forestry changes, such as replacing aspens and poplars with conifers, but has now returned.

CRYPTIC CATERPILLARS

Catacola fraxini eggs overwinter, and the cryptic gray or brown caterpillars develop in spring. They are distinguishable by their flat-bellied appearance, rather like a pipe sliced in half along its length, and are easily mistaken for bark deformities, as they fit into bark crevices, protruding little and remaining largely still. Pupation occurs among leaves pulled together. The pupa is brown, but because of its waxy outer layer appears blue. Adults eclose three to four weeks after pupation.

SUGAR ADDICTS

Adults, which live for around 30 days on average but occasionally much longer, are attracted to sugary substances, such as sap from damaged trees. This is true for all *Catocala* species, of which there are around 250 worldwide.

→ A Blue Underwing drinking sap oozing from an injured tree. It is one of only two *Catocala* species with a blue pattern on its hind wings (the other is the North American White Underwing, *C. relicta*). The Blue Underwing is also the largest member of its genus.

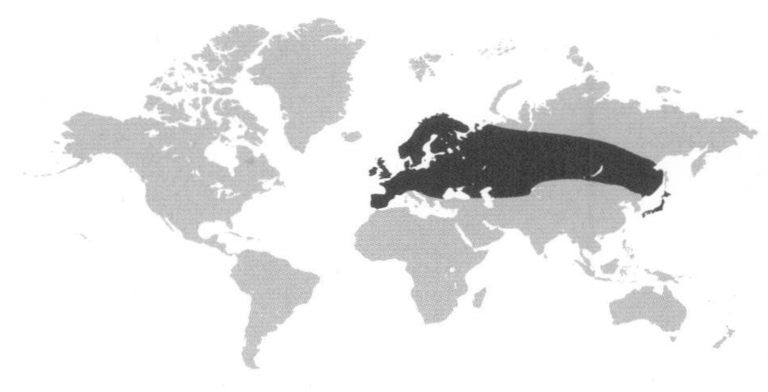

COSSUS COSSUS

Goat Moth

Stout and strong

SCIENTIFIC NAME	*Cossus cossus* (Linnaeus, 1758)
FAMILY	Cossidae
NOTABLE FEATURES	Stout moth with fine, reticulated pattern
WINGSPAN	2¹¹⁄₁₆–3¾ in (68–96 mm)
SIMILAR SPECIES	Certain other cossids, such as *Acossus terebra*

Named for the goat smell of its larvae, *Cossus cossus*, which occurs throughout the Eurasian continent and in northern Africa, is a stout, strong, grayish-brown moth with a hovering flight. It is a highly adaptable species whose larvae feed concealed and protected within the living tissues of trees and vines. The flight season of the non-feeding adults varies with location, but is June to July in the UK.

GOAT SMELL

The female lays batches of several eggs at a time in bark crevices or damaged areas of deciduous host trees, such as willows, poplars, birches, and oaks. The young larvae feed under the bark at first, then bore deep into the wood, developing for four to five years and reaching up to 4 in (100 mm) in length. When mature, they become bright red with a black head. Larvae either leave their tunnels to pupate in the ground, or remain inside their host tree creating a cocoon just beneath its outer surface, so that it is easier for the moth to emerge.

PHEROMONE TRAPS

While forest trees can often withstand the larval damage, a high level of infestation which sometimes occurs, especially in poplar plantations, can destroy them. To combat tree loss from Goat Moth damage, researchers have determined its pheromone composition, so that pheromone traps can be used to monitor populations.

→ While Goat Moth adults may come to light, their huge larvae, which are red with a black head, are only occasionally seen crawling around looking for a place to pupate underground.

SCOLIOPTERYX LIBATRIX

The Herald

Handsome hibernator

SCIENTIFIC NAME	*Scoliopteryx libatrix* (Linnaeus, 1758)
FAMILY	Erebidae
NOTABLE FEATURES	Narrow forewings, purple at base of hind wings
WINGSPAN	1¾ in (44 mm)
SIMILAR SPECIES	*Scoliopteryx aksuana*

Widespread throughout the temperate region of the northern hemisphere, the Herald has characteristic, angled forewings and flies between June and November. When at rest, it mimics a dead leaf. Its caterpillars feed mainly on willows and are bright green, with thin yellow lines running between segments. They pupate in a cocoon wrapped in the leaves of its host.

HIBERNATING IN CAVES

After the first frost, the moths tend to go into hiding in cool, dark subterranean shelters and man-made structures, including caves, mines, cellars, and barns. In caves, they have been found within 13 ft (10 m) of the cave entrance. In warm spring sunlight, they emerge from their overwintering sites, and the next generation returns to shelters in the fall. These moths may also estivate in caves, taking refuge from the summer heat.

LAB RAT MOTH

In one laboratory study, at a temperature of 41°F (5°C)—close to that of their winter cave retreats—the moths remained alive and healthy up to 14 months after capture. Their long dormancy attracted the attention of researchers who studied how their fat body—a specialized organ for storing nutrients, composed of various fat cells—gradually metabolizes lipids and glycogen. Investigating the chemosensory system of these unusual moths, they also discovered a novel compound—(6Z,13)-methylheneicosene—the first example of erebids using this type of chemical as a sex pheromone.

→ The Herald moth is frequently found in caves and mines, where it hibernates from fall until spring.

MOTHS ON CONIFEROUS & WETLANDS PLANTS

Coniferous forests and watery habitats

Coniferous forests and watery habitats might seem unlikely places to harbor a diversity of moths but, wherever there are plants, there are moth caterpillars to feed on them.

MOTHS ON CONIFERS

A range of moth species feed on conifers—the fir, spruce, tamarack, hemlock, larches, and pines that characterize northern forests south of tundra—but are also scattered in every biome. A wealth of others feed on more southerly pines, such as longleaf pines in Florida or Aleppo pine in southern Mediterranean areas, or junipers and cedars in dry habitats, and cypress in swamps. In the southern hemisphere, coniferous moth hosts include *Araucaria* trees, some yew species, and Podocarpaceae such as the pigmy pine or Huon pines.

In the New World alone, the caterpillars of nearly 800 species of moths are known to feed on conifers, more than 500 of which are relatively small moths that specialize on this diet. Some larger moths are also closely connected to conifers, such as the Pine

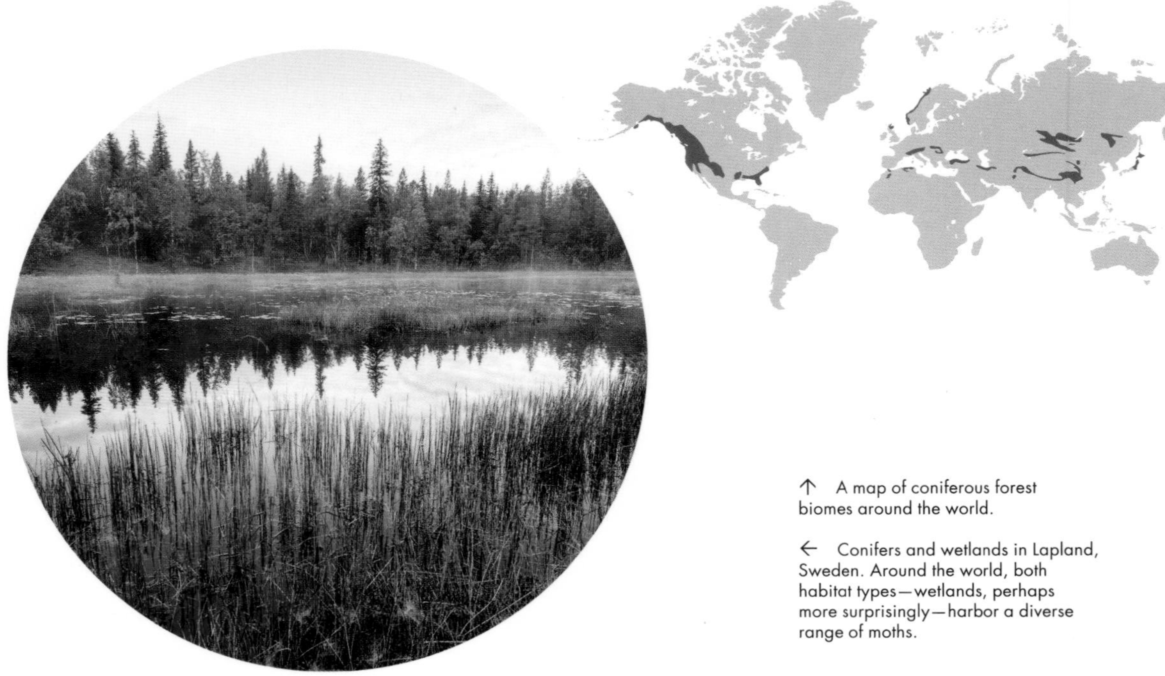

↑ A map of coniferous forest biomes around the world.

← Conifers and wetlands in Lapland, Sweden. Around the world, both habitat types—wetlands, perhaps more surprisingly—harbor a diverse range of moths.

Hawk Moth (*Sphinx pinastri*) in Eurasia, the Pandora Pine Moth (*Coloradia pandora*) in the American Southwest, or *Dirphia araucariae* on *Araucaria* trees in southern South America, and a beautiful day-flying geometrid, *Milionia basalis*, that feeds on *Podocarpus* and *Dacrydium* from Java and Borneo to Japan. Conifers may also be an occasional host for many caterpillars, such as inchworms and cutworms, that feed on them only during outbreaks.

AQUATIC HABITATS

While all conifers have some similarities and are part of the Coniferae plant division, aquatic plants are highly variable. Very different groups of plants can be associated with water and occur at different water depths, with corresponding moths that feed on them. For instance, species such as the Waterlily Borer Moth (*Elophila gyralis*) specialize on floating waterlilies, while others, such as the Australian Pond Moth (*Hygraula nitens*), feed on fully submerged plants, including curly-leaf pondweed, eelgrass, or water thyme (*Hydrilla*). Some moth caterpillars feed on algae on wet rocks, among them a dozen of some 350 *Hyposmocoma* micromoths in Hawaii, whose larvae live inside cases made of silk and debris and, in one species, eat snails.

While some of the wetland moths in and around streams, rivers, and wetlands are fully aquatic, living and breathing underwater, others are semiaquatic and have no special adaptations. In addition, certain species, such as the Pitcher Plant Moth (*Exyra semicrocea*), specialize on plants endemic to wetland habitats or, like the Baldcypress Leaf Roller (*Archips goyerana*), feed on the numerous trees, shrubs, and herbaceous plants rooted in streams, riverbeds, or damp soil. Numerous moth species, including hawk moths, also migrate from forests into wetlands in their search for nectar, adding to the diverse composition of moth communities in aquatic habitats.

↑ The Pine Hawk Moth (*Sphinx pinastri*) caterpillar, whose range extends from northwest Africa to central Siberia, may feed at the top of a tall pine tree, but will descend its long trunk to pupate on the ground in pine-needle litter.

Choosing conifers

The majority of Lepidoptera coevolved with flowering plants, aiding pollination by transferring pollen while feeding on flower nectar. Numerous moth species have since developed the ability to feed on developing cones, bark, or needles of conifers, as specialists or generalists.

OVERCOMING CONIFER DEFENSES

All conifers are well defended by their individual protective plant chemicals and resins. In the Podocarpaceae family, the *Podocarpus* genus alone includes some 100 species of trees and shrubs, yet only six moth species, mostly inchworms (Geometridae), are known to feed on podocarps in South America. Majestic monkey puzzle *Araucaria* trees, once much more widespread in South America, have powerful phenolic compounds that deter many insects from attacking them, while members of the yew family from the Pacific, Australia, and China are similarly defended against major insect damage by alkaloids called taxanes. Feeding on conifers has required special adaptations that have evolved sporadically throughout Lepidoptera, and, as occurs elsewhere, once caterpillars have developed a biochemical ability to overcome a plant's defensive chemicals and even use them for their own protection, they become specialists. Over millions of years of coevolution, this process occurs repeatedly, creating numerous new species.

↖ Some moth species are specially adapted to feed on chemically defended *Araucaria* trees, such as these in the Itaimbezinho Canyon in southeastern Brazil within the Aparados da Serra National Park.

← The Pacific yew or western yew (*Taxus brevifolia*), native to the Pacific Northwest of North America, is also food for moth larvae that can tolerate its chemicals.

SMALL AND SPECIALIST

If they can overcome its defenses, micromoths seem to be more likely than larger moths to specialize on a coniferous plant. They can feed inside a single needle, create a silken web over developing cones and buds, or even tunnel through the cones or bark. In the New World, many of them are moths of the Gelechiidae, Yponomeutidae, and Tortricidae families. Occasional radiations (the rapid evolution of new species) have occurred when one species that has successfully overcome a conifer's defenses then splits into many new species; the development of more than 20 species of tiny *Marmara* gracillariids, close to 80 pine-bark colored *Dioryctria* pyralids, or nine *Eucopina* tortricids with a pine-cone pattern are examples of such speciation events.

Among some of the smallest moths that feed on conifers are *Coccidiphila silvatica* in the family Cosmopterigidae, which have a wingspan of only ¼ in (7–8 mm). Larvae of this moth feed on pines in the Himalayas, but are also carnivorous, eating mealybugs.

← The 79 species of *Dioryctria* snout moths are hard to tell apart. In the United Kingdom, *D. abietella* and *D. simplicella* feed as caterpillars on pine cones, shoots, and buds.

THE FEARED SPRUCE BUDWORM

The Eastern and Western Spruce Budworms (*Choristoneura fumiferana*, right, and *C. freemani*), members of a large worldwide genus of more than 40 species, are major and highly costly pests on spruce and balsam fir trees, preferring flowering trees, whose buds are rich in pollen proteins. Scientists can now determine the size of centuries-old outbreaks by studying the sediment of nearby lakes, where the wing scales of moths are well preserved. Bird predation usually prevents population explosions of the Spruce Budworm, as species such as Canada warblers and golden crowned kinglets consume more than 80 percent of the moth's eggs and larvae. Parasitoids also contribute to population control; a 2020 study identified nine fly species and 27 of parasitic wasps that attack Spruce Budworm larvae, indicating the importance of healthy biodiversity for balancing pest populations in any given ecosystem.

CARPENTERS AND WOOD BORERS

Most moths cannot feed inside living conifer wood, partly because of its resins and other powerful chemicals. The few wood-boring moth caterpillars that have evolved the ability to overcome these tree defenses include the Pine Carpenterworm (*Givira lotta*), which mines only the pine bark, and the Pitch Mass Borer (*Synanthedon pini*), in the clearwing moth family Sesiidae, which bores into the trunk.

DIVIDING THE SPOILS

Caterpillars of the Pine Tube Moth (*Argyrotaenia pinatubana*), as its name suggests, make tubes out of needles by hollowing them, preferring white pine in the northeastern United States and Canada. Another tortricid, the Juniper Budworm (*Choristoneura houstonana*), creates silken tubes and feeds only on the foliage of the Ashe juniper (*Juniperus ashei*). Throughout Eurasia, outbreaks of the Larch Bud Moth (*Zeiraphera griseana*) can significantly damage various coniferous trees as the caterpillars move around, attacking numerous needles, and leaving webs and frass behind. The outbreaks are cyclical in nature; populations increase periodically, but then decrease as a result of viruses, other natural enemies, or a reduction in available food. The caterpillars of the Eurasian tortricid moth *Cydia duplicana* mine the bark of pines, but only near damaged bark, or where fungus has attacked it, presumably obtaining nutrients from chemicals that result from an injury, rather than from the bark itself.

LIVING OUT OF A BAG

One of the oddest and best camouflaged caterpillars is the Evergreen Bagworm (*Thyridopteryx ephemeraeformis*), found in the eastern United States. Its polyphagous caterpillars live inside little dwellings they construct with a diverse range of twigs and greenery from their many host plants. Juniper is one of the bagworm's favorites and is frequently attacked even in unlikely habitats such as cityscapes where these evergreen cone-shaped trees are used as ornamentals. Decorated with pieces of plant material, bags with larvae inside can grow up to 2 in (60 mm) in length. Larvae can close their bag when sensing danger, largely avoiding parasitoids and predators. Males eclose from their protective homes and search for the grub-like females, which stay inside the bag and emit pheromones.

The Evergreen Bagworm is the best known of its family Psychidae that includes more than 1,300 moths worldwide, many of which are small and rare. One similar Australian species is the Faggot Case Moth (*Clania ignobilis*), whose larvae attack *Callitris* cypress pines.

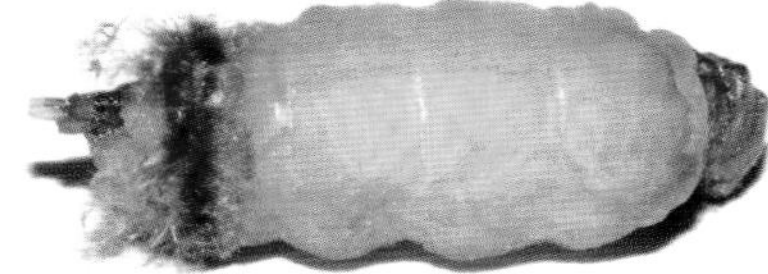

FEMALE

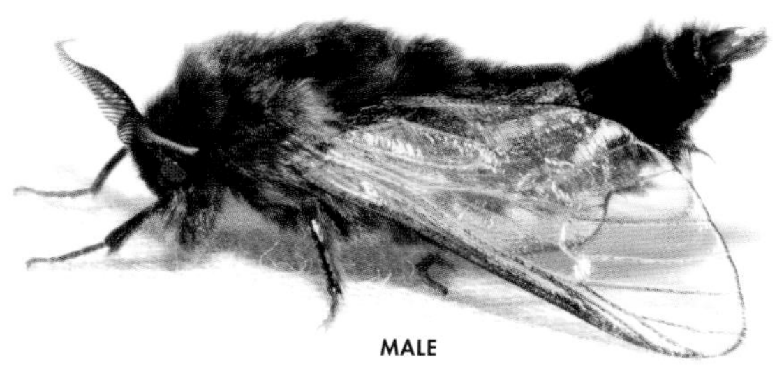

MALE

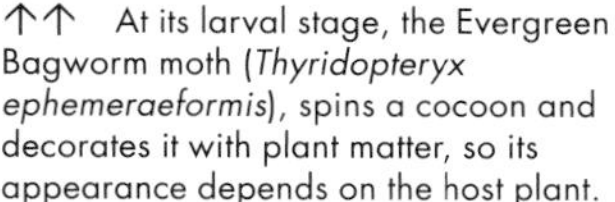

↑↑ At its larval stage, the Evergreen Bagworm moth (*Thyridopteryx ephemeraeformis*), spins a cocoon and decorates it with plant matter, so its appearance depends on the host plant.

→ Adult male Evergreen Bagworm moths resemble bumblebees, while females are grub-like, with no wings or other appendages.

← The Larch Bud Moth or Larch Tortrix (*Zeiraphera griseana*) can develop on a variety of conifers.

← The Larch Casebearer (*Coleophora laricella*) caterpillar lives inside a case made of hollowed out host plant needles.

→ Its coloring and shape make this Australian moth, *Meyrickella torquesauria*, cryptic on its host plant *Callitris columellaris*, known as the white cypress pine.

↓ *Tracholena sulfurosa* is one of several tortricid species in Australia that are able to overcome the defenses of conifers and may tunnel into the bark of *Callitris* trees.

MINING CONIFER NEEDLES

The Larch Casebearer (*Coleophora laricella*) belongs to a genus that includes more than 1,000 tiny moths and comprises 95 percent of the Coleophoridae family. The newborn *C. laricella* caterpillar penetrates a larch or tamarack needle via the bottom of its egg, bypassing any outside dangers, and mines inside the needle for a couple of instars. When larger, the caterpillar makes a case of silk and empty needle shells. The Larch Casebearer was introduced to the United States from Europe in the late nineteenth century and has since spread to many larch stands, frequently damaging and weakening trees, making them vulnerable to bark-beetle infestations.

ANTIPODES OF PINES

In Australia, and surrounding islands, all 16 species of *Callitris* trees, which are members of the cypress family and endemic to the Antipodes, occupy a niche similar to that of conifers of the northern hemisphere. Known as cypress pines for their intermediate appearance, in Australia alone, they occupy nearly 5 million acres (2 million ha).

Callitris are hosts to many moths, including a recently discovered species that is so unusual that it has become the basis for an entirely new moth family. First described in 2015, the Enigma Moth (*Aenigmatinea glatzella*), which was found on Kangaroo Island off the coast of South Australia, is a primitive species with highly reduced mouthparts. As a caterpillar it feeds on Murray cypress pine (*Callitris gracilis*), mining its bark. The Australian tortricid *Tracholena sulfurosa* burrows inside the bark of various conifers including *Callitris*. The peculiar-looking, rusty brown, white, and black patterned noctuid *Meyrickella torquesauria* feeds on the white cypress pine (*C. columellaris*).

MOTHS ON MONKEY PUZZLE TREES

As caterpillars, several moths feed on *Araucaria* trees, which can reach a height of 250 ft (76 m) and grow in and around Australia as well as South American countries, including Chile, where the monkey puzzle tree (*A. araucana*) is the national symbol.

Araucarivora gentilii, found in Argentina and Brazil, is the only member of its genus and part of the ancient family of grass-miner moths Elachistidae. As its genus name suggests, the caterpillar of this tiny moth feeds only on *Araucaria* trees and is so small that its entire life cycle, including pupation, occurs within one needle in the mine the larva makes. The caterpillar of the silk moth *Dirphia araucariae*, which is thousands of times larger, eats the needles of the candelabra tree (*Araucaria angustifolia*), where it is perfectly camouflaged, as its spines and striped color pattern imitate its host plant's needles.

↖ One of 47 species in its genus, *Dirphia araucariae*, a saturniid moth, feeds on needles of a candelabra tree (*Araucaria angustifolia*) in Brazil.

← The cones of living *Araucaria* trees bear a strong resemblance to fossilized cones dated to the mid-Jurassic period, indicating the ancient origins of this tree genus.

Silk moths on conifers

A number of spectacular large moths in the Bombycoidea family of silk moths, hawk moths, and relatives specialize on conifers. For example, many of the 14 species in the saturniid genus *Coloradia* or 27 species in the hawk moth genus *Sphinx* are associated with pines but have distinctive geographic and ecological preferences. Their larvae are usually striated green, camouflaging them among needles, while the adults' forewing colors tend to blend with pine bark.

SILK MOTHS ON LARCH AND PINES

The Columbia Silk Moth (*Hyalophora columbia*) occurs between July and September from New York State to northeast Canada, and is known as the Larch Silkworm for its caterpillar's diet of the deciduous conifer, American larch (*Larix laricina*). The moth, which has a large wingspan of almost 4 in (100 mm), has a very similar subspecies, Glover's Silk Moth (*H. columbia gloveri*), whose caterpillars feed elsewhere on various non-coniferous plants but have trouble developing on larch. This may be a sign of a larch diet causing speciation (the formation of a new separate species), prompted by *H. columbia's* choices of host plant.

The Pine Devil (*Citheronia sepulcralis*), which got its name for the hornlike projections on its big, gray-brown caterpillars, feeds on pine needles. Unlike its better-known relative, the Royal Walnut Moth (*Citheronia regalis*), which feeds on broadleaf trees, the Pine Devil caterpillar is not intimidating and relies on crypsis at all stages of its life.

← This caterpillar of the Columbia Silk Moth (*Hyalophora columbia*) from the Magdalena Mountains, New Mexico, belongs to the western North American subspecies that can eat leaves of deciduous plants. The almost identical subspecies in the eastern United States feeds on conifers.

→ Cryptic caterpillars of the Pine Devil moth (*Citheronia sepulcralis*) feed on various pine species in the eastern United States, mostly in coastal habitats. Females lay two to three eggs at a time, and the larvae hatch a week later. At maturity, when they reach a length of around 4 in (100 mm), they pupate underground.

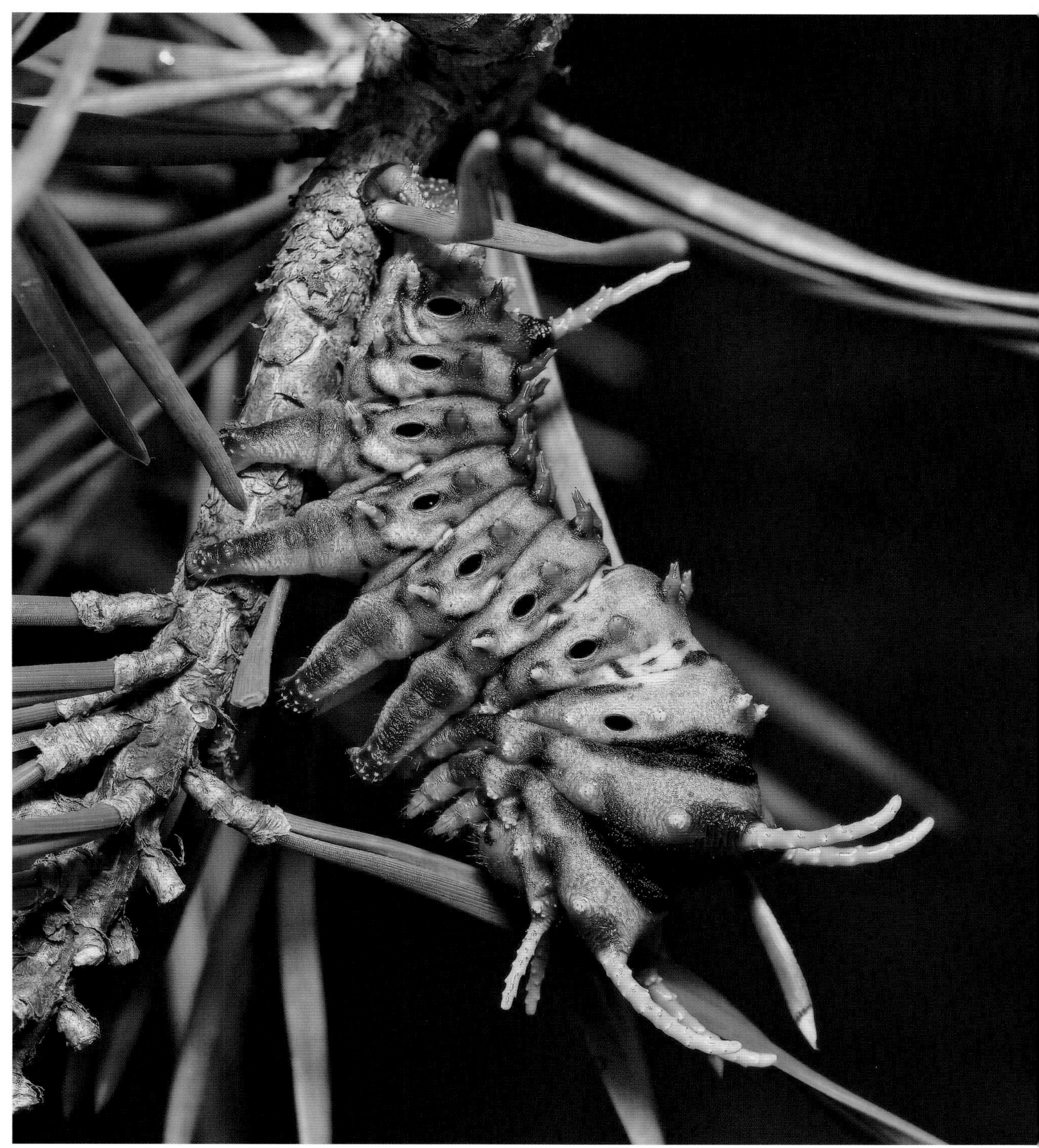

CATERPILLAR FEAST

For centuries, caterpillars were an important and sustainable food staple for indigenous people in California, who searched them out much as entomologists do—by looking for frass (larval excrement) under a host tree. They used to visit tall Jeffrey pine trees in mid-June and, if they saw frass, returned when the large larvae of the Pandora Pine Moth (*Coloradia pandora*) were mature, making trenches around the trees to collect the caterpillars as they crawled down to pupate. According to one 1921 account of the practice, 1½ tons (1.4 tonnes) of caterpillars were collected by a single extended family during a "good caterpillar year" (their term for an outbreak), which occurs every 20 to 30 years. The caterpillars, which are rich in protein and fats, were cooked in sand piles, preheated with burning wood, then stored for up to two years before they were boiled and eaten.

Pandora Pine Moths develop over two years, overwintering in groups as partially grown caterpillars, while the closely related Lusk's Pine Moth (*C. luski*), found in eastern California, overwinters as a pupa after just one year of feeding. A black-winged species of *Coloradia*, the Prchal's Pine Moth (*C. prchali*), first described in 1992 in northern Mexico, feeds on a variety of pines and junipers in the Sierra Madre Occidental pine–oak forest.

↑ ← The Pandora Pine Moth (*Coloradia pandora*) and its caterpillar. The moth is found from California east to Texas and north to Washington State. Its large larvae, which feed on Jeffrey pine in California, are so numerous that for centuries, they were an important food supplement for local Native American tribes.

OTHER BOMBYCOIDS ON PINES

Among other bombycoid species that feed as caterpillars on pines and spruce is the Pine-tree Lappet (*Dendrolimus pini*) of Eurasia, which, during occasional outbreaks, can strip coniferous forests of their needles. The moth is furry and gray, with chocolate-brown patterned forewings. The female lays a pile of eggs on the bark of host trees, and the caterpillars first feed together but then spread out, consuming needles. In the northeastern United States and across Canada, caterpillars of the Larch Tolype (*Tolype laricis*), a black-and-white furry moth, feed on a variety of conifers, and mature and fly in late summer.

↑ The Pine-tree Lappet (*Dendrolimus pini*) caterpillar frequently defoliates conifers in some parts of Europe and Asia.

↖ Traditional Japanese straw belts, called *komomaki*, are wrapped around pine trees to trap pine-feeding pests and encourage them to pupate in the straw, rather than crawling down the trunk to pupate underground. Where this is used as a control method, as here, the belts are then burned in spring before the adults eclose. Similar methods are also used by entomologists eager to obtain pine sphinx moth specimens.

Inching and looping through the needles

In the northeastern United States, in and around the New England area, dozens of species of inchworms and noctuid moths have been recorded on conifers—some restricted to the pine and cypress family plants, and some polyphagous, feeding only occasionally on conifers. Those that specialize on conifers tend to have a white-striped green coloration that perfectly matches their host plants, while those with a broader diet often resemble sticks.

GEOMETRIDS ON CONIFERS

Many of the 23,000 species in the family Geometridae are highly polyphagous, and some feed on conifers among other plants, while others have become specialists. In eastern North America, they include certain emerald moths (*Nemoria* spp.), named for their green coloring, or the bark-like patterned *Hydriomena* spp., and pugs (*Eupithecia* spp.). Caterpillars of the Morrison's Pero Moth (*Pero morrisonaria*) eat a variety of firs and some broad-leaved trees. As an adult, this moth mimics a dry leaf, while the caterpillars resemble twigs, and their appearance varies in a bid to fool birds that develop search patterns for their prey. The Esther Moth (*Hypagyrtis esther*) that feeds on pines as a caterpillar, is an equally convincing dry-leaf mimic as an adult, and almost impossible to spot when it rests on the ground. *Iridopsis cypressaria* is one of numerous white, flat-winged moths whose coloration blends with that of lichens on tree bark. Unlike the other 20 members of its genus, however, it specializes on conifers.

→ The noctuid Pine Beauty (*Panolis flammea*), whose caterpillar feeds on conifers, is widespread throughout Eurasia.

↓ The Esther Moth (*Hypagyrtis esther*) caterpillar is one of 165 species of inchworm (geometrid moth larvae) in the New World that are known to feed on conifers; 70 of which are specialists.

PROCESSIONS AMONG PINES

Like the larvae of other processionary moths, discussed in previous chapters (see page 197), larvae of the Pine Processionary (*Thaumetopoea pityocampa*) of southern Europe, move around nose-to-tail in large groups. They live communally on pine trees, feeding on needles, and thermoregulate their silken nests by strategically positioning them where they can be warm even on colder days. Among other adaptations to group living, they lay down a chemical trail when foraging to enable them to find their way back to the nest. Pine Processionary caterpillars can defoliate pines and also, occasionally, larches in southern Europe.

→ A nest of Pine Processionary (*Thaumetopoea pityocampa*) caterpillars.

OWLETS AND RELATED MOTH GROUPS

With over 12,000 species, owlet moths and their relatives are so numerous and diverse that taxonomists are almost constantly reclassifying them into different families and subfamilies. Most of them are small, brown, fast-flying moths, and their caterpillars are often smooth and are frequently known as cutworms. The Pine Beauty (*Panolis flammea*), for example, that flies from Spain to the Arctic Circle, has caterpillars with green stripes to camouflage them effectively against coniferous foliage. This small moth is on the wing in the spring, so that its caterpillars, which feed in groups, can take advantage of fresh needles on pines and young leaves of other trees. From southern Canada to New Mexico, another noctuid, the Abstruse False Looper (*Syngrapha abstrusa*), feeds on spruce and jack pine, while the related Salt and Pepper Looper (*S. rectangula*) favors Douglas and balsam fir, hemlock, and spruce as host plants.

MOTHS ON FERNS, LICHENS, AND CYCADS

Ferns, lichens (symbiotic organisms formed by fungus and algae), and cycads appeared on Earth long before conifers and angiosperms. Thus, it is not surprising that some moths have adapted to feeding on them as caterpillars, despite their frequently toxic nature.

Lichen moths

The caterpillars of more than 2,700 species of moth in the tribe Lithosiini in the arctiine subfamily of Erebidae, frequently thrive on a diet of lichens, which grow on tree bark, rocks, and stones. Lichens produce defensive phenolic compounds, which caterpillars feeding on them can accumulate, and which may also provide some chemical protection and novel pheromones for these colorful moths. In addition to *Miltochrista* and *Hypoprepia* shown here, see also the account of *Barsine orientalis* (page106).

Cycad moths

The Echo Moth (*Seirarctia echo*) is a rare example of a moth whose caterpillar feeds on cycads—primitive gymnosperm plants related to conifers, but far more ancient. Cycads contain potent protective cyanogenic compounds, so few insects can feed on them, but in Florida, the Echo Moth has developed the ability to detoxify them and feeds on coontie and other *Zamia* spp.

Fern moths

The Florida fern caterpillar (*Callopistria floridensis*) is one of a number of moths that have adapted to feeding on ferns, primitive plants that reproduce via spores.

↗ *Miltochrista pulchra*—one of more than 50 species in its genus—is found from the Russian Far East and Japan to Yunnan province in China and the Korean Peninsula. It is one of many species known as lichen moths for the food preference of their larvae, or footmen for the stance of the adults at rest.

→ The Scarlet Lichen Moth (*Hypoprepia miniata*) is a member of a complex of similar *Hypoprepia* species found in North America. Bright colors are indicative of the defensive compounds their caterpillars acquire from lichens (mostly from those that grow on pines) and pass on to the adults.

← The Echo Moth (*Seirarctia echo*) caterpillar feeding on coontie. Found in the eastern United States, this moth is the only member of its genus. Its caterpillars can develop on other plants but prefer coontie—a popular ornamental plant in Florida; the Echo Moth is among very few species that can detoxify cycasin—the toxic glucoside found in coontie and other cycads.

↓ The Florida Fern Moth (*Callopistria floridensis*) caterpillar is one of several moth species that can overcome the chemical defenses of ferns, which include compounds such as proanthocyanidins (condensed tannins).

Moths living in and around water

The composition of moth communities in wet habitats is quite complex, as the multitude of species living in or around ponds, streams, rivers, and wetlands have quite differing needs. Truly aquatic moths, with larvae adapted to breathing underwater, are a tiny minority, though still represented by more than 800 known species across the world, and probably many more whose biology remains to be described. Regardless of how these moths utilize wetlands plants, all play important roles in their respective ecosystems, and without them the bird species that thrive in wetlands and attract so many nature-loving visitors would greatly suffer.

UNDERWATER MOTHS

The Water Veneer (*Acentria ephemerella*), native to Europe and also established in North America, not only has aquatic caterpillars; most of its females also live underwater, rising to the surface only to mate. In most aquatic moth species, however, the adult moths of both sexes are terrestrial, while their larvae develop underwater. Certain aquatic moth caterpillars, such as those of the Australian Pond Moth (*Hygraula nitens*), breathe through gills—hairlike filaments that are extensions of their tracheal system.

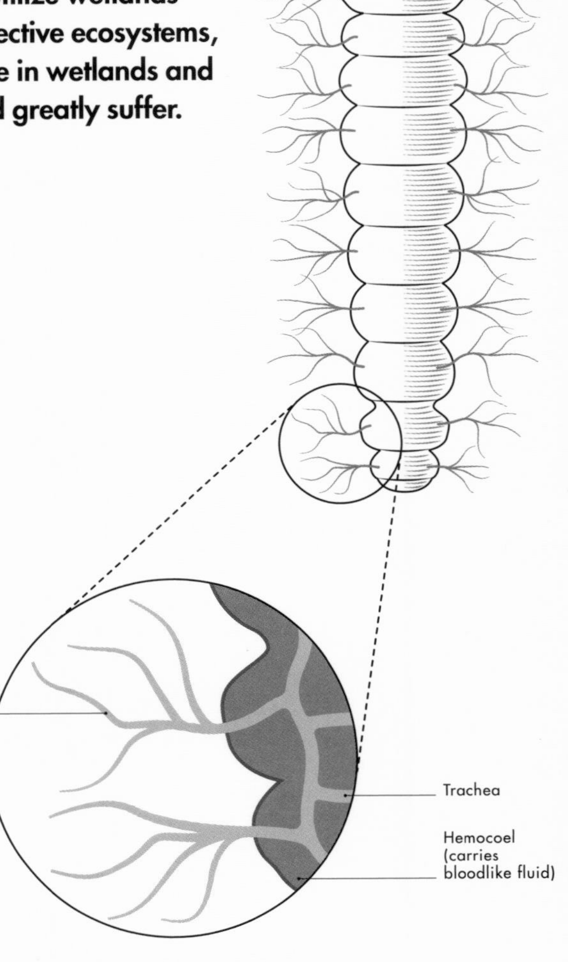

Aquatic species equipped with gills

Some aquatic moth species, such as the Australian Pond Moth (*Hygraula nitens*) have gills—branching tracheal outgrowths that absorb oxygen from water and transport it to the hemolymph.

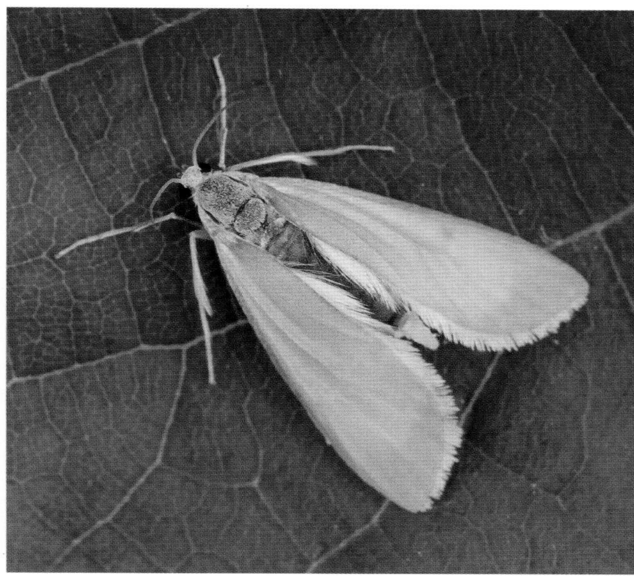

In North America, males of the Polymorphic Pondweed Moth (*Parapoynx maculalis*) are snowy-white with black patterning, while the females are a drabber gray. Its genus *Parapoynx* has almost 60 described species, most of which have aquatic larvae that have developed gills and feed on a variety of aquatic plants, with a preference for waterlilies. In Europe, the similar Ringed China-mark (*P. stratiotata*) has been reared on 11 different aquatic plant species, and flies from May to September (or June to August in southern Britain).

OTHER AQUATIC STRATEGIES

Some species construct tubes of leaves that both create an air bubble around them and conceal them from predators. Semiaquatic caterpillars of the genus *Epimartyria*, small primitive metallic moths, have tiny, elevated dots (micropapillae) on their skin, which hold a thin layer of air on their outer surface, enabling them to spend short periods in water-saturated areas of their natural habitat of North American woodland swamps and ditches. Caterpillars of *Hyposmocoma* moths can live on land and also in well-oxygenated streams, adhering to rocks, but may die in still water. It is thought that they can breathe through their skin. Larvae in this genus, which has diversified into more than 350 species in the ancient Hawaiian chain of volcanic islands, live inside purse-shaped structures decorated with tiny particles of sand held together by silk. A dozen or so of the species have larvae that can feed on algae growing on wet rocks underwater, but readily complete their development above the surface if water levels fall. Remarkably, one the *Hyposmocoma* species (*H. molluscivora*) on the islands of Maui and Molokai, has added mollusks to its diet, becoming a predatory carnivore.

↖ The Water Veneer (*Acentria ephemerella*) caterpillar breathes by absorbing oxygen diffused in water directly through the cuticle.

↑ The male of the Water Veneer moth is terrestrial and has wings (as here), while the wingless females are aquatic.

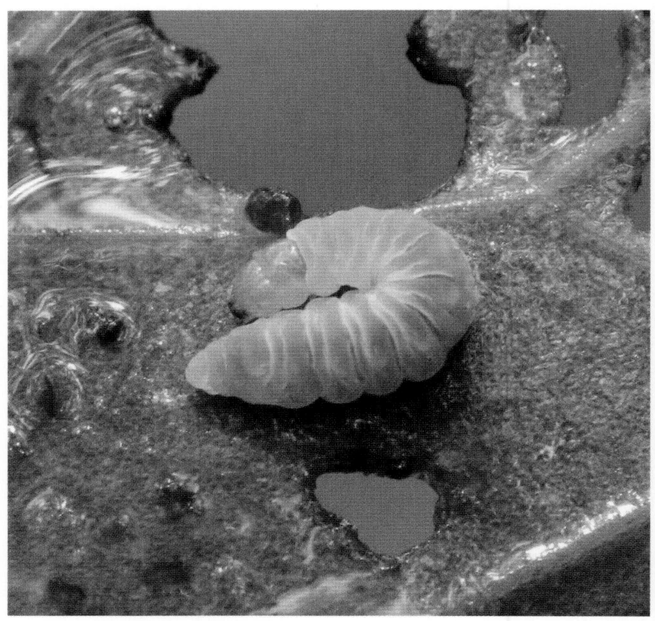

→ Waterlily Borer (*Elophila gyralis*) caterpillars make holes in waterlily leaves as they feed on them. They also create purselike shelters out of leaf pieces and swim within them, keeping a good supply of oxygen inside their home.

↓ Waterlily Borer Moth adults are quite variable, ranging in forewing coloration from a patterned chocolate shade (as here) to orange or gray with very little patterning.

MOVING THROUGH WATER

While the aquatic female Water Veneer moth propels herself through water using the long-haired fringes on her legs as oars, neither aquatic nor semiaquatic caterpillars have developed any special adaptations to help them paddle or swim; most move around much as terrestrial species do, using thoracic legs and prolegs with crochets, and are usually found on the plants they feed on. Winged adults of aquatic species can swim briefly after they eclose, but must then dry their wings before flying. Some aquatic moth females are wingless; winged males fly in and mate with them.

SEEKING OUT AQUATIC PLANTS

While the portion of the moth fauna considered truly aquatic is relatively small, the number of moths associated with aquatic host plants is much greater. Ultimately, it is not the water that attracts moths but the vast plant resources. The Waterlily Borer Moth (*Elophila gyralis*) has aquatic larvae that feed on waterlily leaves. Found in the same habitat, caterpillars of the related Waterlily Leafcutter Moth (*E. obliteralis*) feed

on more than 60 different aquatic plant species, from hydrilla (*Hydrilla verticillata*) to water lettuce (*Pistia stratiotes*). The Australian Pond Moth caterpillar eats many aquatic plants, including various pondweeds and non-native hydrilla species. Aquatic caterpillars of moths in the genus *Petrophila* live within a silken web in streams and feed on algae off the rocks. The adults of these species can be found on flowers close to bodies of water and sport a jumping spider-like pattern on their hind wings.

↑ Several moth species have adapted to navigate the perils of living inside Pitcher Plants or Trumpet Pitchers (*Sarracenia* spp.), such as these, feeding on them as caterpillars.

↗ Pitcher Plant Moths (*Exyra semicrocea*), found from Florida to North Carolina and westward to Texas, spend days inside their host plant.

EATING THE CARNIVORE

The Pitcher Plant Moth (*Exyra semicrocea*) is one of three intrepid North American moth species whose larvae feed on carnivorous trumpet pitcher plants (*Sarracenia*). The plants, which grow in bogs, fens, and wet grasslands in Texas, the Eastern Seaboard, and the Great Lakes area, attract insects into their pitcher-shaped leaves, where powerful enzymes digest their prey. The caterpillars walk inside the narrow funnel but are prevented from falling too far down by lappets—specialized projections equipped with long stiff setae. The larvae also lay a silk thread, which they can use to pull themselves back up the funnel. They spend the winter (when the plant is not producing enzymes) under a dry pile of dead prey remains inside the pitcher, and pupate in a fresh, untouched pitcher plant, burrowing into the stem and constructing a loose cocoon. Adults can spend days inside the plant thanks to a specialized structure of pretarsal claws that enable them to cling on. The Pitcher Plant Moth's range has contracted because the bog area where pitcher plants grow has reduced; large parts of northern central Florida no longer contain this once common habitat type.

Moths in wetlands

In addition to being food for fish, frogs, birds, bats, and aquatic predaceous insects, such as water bugs and dragonflies, many moths are important and sometimes unique pollinators of flowers that grow in wetlands. Whole communities can be associated with these ecosystems, including passing moths attracted by nectar when flowers are in bloom.

ALONG THE WATER

A number of moths specialize on the many herbaceous plants, grasses, and trees that are rooted in swamps or in water-inundated coastal sands, sometimes forming dense stands. Several moth species use the vast expanses of cordgrass along the coast of the Atlantic Ocean, including the Two-striped Cordgrass Moth (*Macrochilo bivittata*) and Louisiana-eyed Silk Moth (*Automeris louisiana*). In the swamps of northern North America, caterpillars of the Canadian Sphinx (*Sphinx canadensis*), feed on black ash trees, while adults may migrate far and wide in search of nectar sources.

← The Banded Sphinx (*Eumorpha fasciatus*), pictured here in a Costa Rican mangrove swamp. Its caterpillars favor wetland plants such as primrose willows.

↗ A ghost orchid flower in Fakahatchee Swamp, southern Florida.

→ The Fig Sphinx (*Pachylia ficus)* is a hawk moth that was observed visiting ghost orchids in south Florida. Its proboscis is long enough to reach deep into the flower to obtain nectar; orchid pollinia then adhere to its head or thorax.

ORCHID NECTAR THIEVES

One illustration of a moth's importance in the wetland ecosystem involves the beautiful and endangered ghost orchid (*Dendrophylax lindenii*), made famous by the 1994 book and 2002 movie *The Orchid Thief*. The flowers of this ephemeral species open, sometimes for a single night, in Florida's Fakahatchee Swamp, often far above the water, attached to a cypress tree. The orchid is pollinated by hawk moths, whose long proboscis can reach into the flower and obtain nectar from its spur. Recent photographic evidence from trap cameras suggests that the pollinia from the orchid adheres to the head or thorax of these moths, from where it can be carried to another orchid.

FRED THE THREAD

New Zealand's peat bogs are home to what is quite probably the world's thinnest moth caterpillar—*Houdinia flexilissima*, also known as Fred the Thread, first discovered in 2006, and placed in the leaf-mining family Batrachedridae. The caterpillar, which is approximately ¾ in (20 mm) long when fully grown, reaches only 1 mm in width, and lives within the stem of the giant wire rush (*Sporodanthus ferrugineus*). Its species name *flexilissima* reflects its remarkable ability to maneuver its body through the plant stem as it eats. Like other members of its leaf-mining Batachedridae family, the brown adult moth is also tiny; its delicate, feathery wings span only ½ in (12 mm).

SURVEYS OF WETLANDS FOR MOTHS

Several past and ongoing surveys of moths in wetlands around the world have demonstrated the importance of these ecosystems for moth conservation. For example, one three-year survey in the damp Motovun Forest along Croatia's Mirna River and its tributaries identified more than 400 moth species. An ongoing survey of Paynes Prairie, a large freshwater marsh in Florida, has to date yielded over 1,000 moth species.

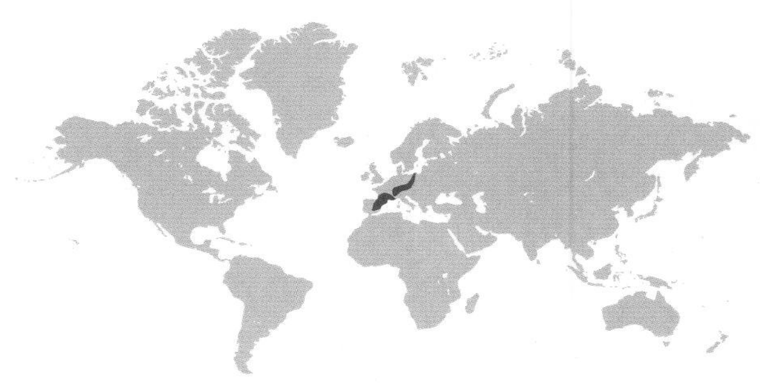

GRAELLSIA ISABELLAE

Spanish Moon Moth

Wings like a stained glass window

SCIENTIFIC NAME	*Graellsia isabellae* (Graëlls, 1849)
FAMILY	Saturniidae
NOTABLE FEATURES	Venation with corresponding stripes on otherwise green wings, hind wings with tails
WINGSPAN	2⁹⁄₁₆–4 in (65–100 mm)
SIMILAR SPECIES	The Indian Moon Moth (*Actias selene*) and other *Actias* spp.

Also known as the Stained Glass Moth for the colored, frame-like venation of its wings, the Spanish Moon Moth, which flies in Spain and France, is one of Europe's most exotic species and has featured in several films, including Philippe Muyl's *Le Papillon* (2002). It occurs at elevations of between 500 and 1,800 ft (152 and 550 m) in the French Alps and Pyrenees, and in other Spanish mountain ranges.

LIFE CYCLE

The silk moth female lays up to 150 eggs, a few at a time, on the pine trees *Pinus sylvestris* and *P. nigra*. The caterpillars, which are initially gray brown and twiglike when young, consume tough mature needles, and become cryptically green and brown as they steadily mature over a period of up to two months. They pupate in a thin, golden-brown cocoon at the base of the tree or in leaf litter.

GENETICALLY UNIQUE POPULATIONS

A 2016 DNA study demonstrated that the French Alps population of the Spanish Moon Moth is more closely related to those in the Pyrenees, than to other Spanish mountain populations, but that each population has a separate, unique genetic composition, demonstrating the importance of conserving as many populations as possible of any one species.

Graellsia isabellae is thought to be an early offshoot of the evolutionary tree branch that also includes the comet moths of Africa and *Actias* moths of Eurasia and North America, such as the Luna (*A. luna*) and Indian Moon Moth (*A. selene*). *Graellsia isabellae* has been hybridized with *A. selene* in captivity—a clear indication of their close relationship, supporting the view of some that they should be in the same genus *Actias*.

→ Monitoring of the Spanish Moon Moth, prompted by conservation concerns, indicates considerable genetic differences in its geographically separated populations.

Needle nibbler
The caterpillar of the Spanish Moon Moth feeds on pine needles, preferring native species.

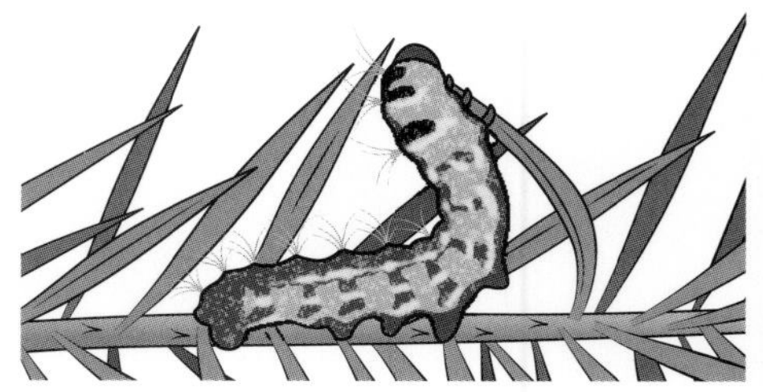

ELOPHILA NYMPHAEATA

Brown China Mark

Finely patterned beauty

SCIENTIFIC NAME	*Elophila nymphaeata* (Linnaeus, 1758)
FAMILY	Crambidae
NOTABLE FEATURES	Small shiny moth, with highly patterned wings
WINGSPAN	⅝–¾ in (16–20 mm)
SIMILAR SPECIES	Superficially resembles several non-aquatic crambids, such as the Beautiful China Mark (*Nymphula nitidulata*) and Variegated Pearl (*Synclera traducalis*)

Found from Europe across Asia, north of the Himalayas to China, the tawny brown-and-white patterned Brown China Mark moth is widespread and fairly common around wetlands, flying throughout the summer months.

AQUATIC HABITS

While this and other *Elophila* species may appear similar to other small moths of the crambid family, what sets them apart is the aquatic habits of their caterpillars. The light-brown larvae make flat oval cases from two pieces of leaves cut from their host plants—pondweed and waterlilies—which they tie together to create a purselike shelter. Each caterpillar floats inside its case, moving between host plants and constructing ever larger shelters as it grows. Larvae that hatch in the fall may also make cases from duckweed leaves, and more mature larvae occasionally mine pieces of bulrush floating on the water surface.

BREATHING UNDERWATER

In some *Elophila* species, including the Brown China Mark, their aquatic lifestyle led to the development of tracheal gills—extensions of the trachea outside the body that help the caterpillar acquire oxygen from the water. The larvae diapause in the winter beneath the surface in their water-filled cases, breathing through their gills. In the spring, they rise to the surface and fill their case with air again. When preparing for pupation, the caterpillar attaches its final case to a stem of a water plant 2–4 in (50–100 mm) beneath the surface and makes a hole in the stem to obtain oxygen from the plant when it goes into the pupal stage.

→ The Brown China Mark moth's hind wings (hidden here) are similarly patterned to the forewings but have more pronounced wavy bands of chocolate brown or black.

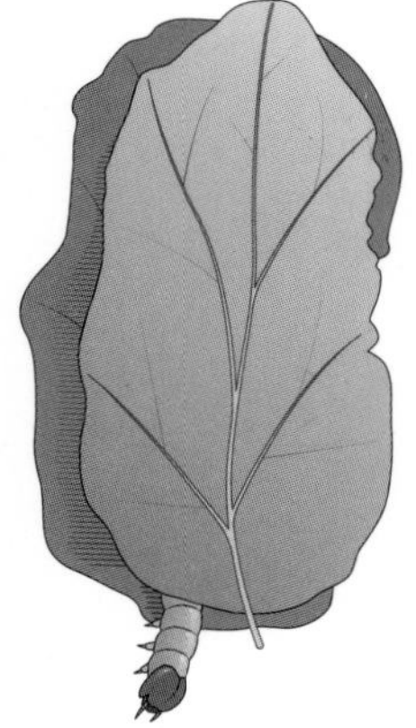

Leafy shelter

Caterpillars of the Brown China Mark may make purselike shelters out of waterlily leaf pieces that they cut to size and shape, and stick together with silk. Air is trapped within their shelters, allowing them to spend time underwater.

PANOLIS FLAMMEA

Pine Beauty

Patterned pine feeder

SCIENTIFIC NAME	*Panolis flammea* (Denis & Schiffermüller, 1775)
FAMILY	Noctuidae
NOTABLE FEATURES	Two large cream-colored spots on each forewing
WINGSPAN	1¼ – 1$\frac{9}{16}$ in (32–40 mm)
SIMILAR SPECIES	Other *Panolis* spp., such as *P. variegatoides* (Southeast Asia)

A common species of pine woods in Europe, the Pine Beauty is also found in Asia Minor, as far east as western Siberia and Asia Minor, and north to the Arctic Circle. Its attractive, highly patterned forewings range from rich orange-brown to gray, while its hind wings are brown.

HOST PLANTS

The Scots pine and lodgepole pine, as well as fir, cypress, and juniper are among its host plants. The species forms one generation per year, flying in April to June. The young larvae feed on new foliage; overwintering pupae spend the winter in needle litter under the trees.

VARIABLE NUMBERS AND LONGEVITY

Females can lay anything from 90 to 300 eggs. They lay large numbers of smaller eggs on more nutritious host plants where their young develop faster; they also live longer as adults. Though the moth is most common in regions where the average temperature is 46°F (8°C), numbers tend to increase during hot, dry summers, which can lead to later population explosions.

FRIEND OR FOE

As elsewhere, outbreaks are part of the natural cycle and often followed by a crash when numbers of the moths' natural enemies increase in response. In natural pine forest ecosystems, the Pine Beauty is an important part of the food web, providing nutrition for birds and other insects. However, the species is considered a pest on lodgepole pine plantations. As a result, its pheromones, identified as (Z)-9-tetradecenyl acetate, (Z)-11-hexadecenyl acetate, and (Z)-11-tetradecenyl acetate in the ratio 100:5:1 have been synthesized, and males are captured in pheromone traps for monitoring and to determine control measures.

→ The Pine Beauty can increase dramatically in numbers after hot dry summers and defoliate pine forests, but natural enemies, such as parasitic wasps and predatory birds, soon reduce populations to their normal size.

SPHINX PINASTRI

Pine Hawk Moth

Master of disguise

SCIENTIFIC NAME	*Sphinx pinastri* (Linnaeus, 1758)
FAMILY	Sphingidae
NOTABLE FEATURES	Mouse-gray with black dashes, long wings typical of Sphingidae
WINGSPAN	2¾ –3¾ in (70–96 mm)
SIMILAR SPECIES	Other *Sphinx* species, such as the Canadian Sphinx (*S. caligineus*) and Asian Pine Hawk Moth (*S. morio*), and the Northern Pine Sphinx (*Lapara bombycoides*) in North America

Found in the open pine forests of much of Europe and northwest Asia, Pine Hawk Moths, like many other members of their genus, have a pattern of vertical black dashes on their narrow gray wings, which provides excellent camouflage against conifer bark.

POWERFUL POLLINATORS

The adults sip the nectar of evening primrose and other flowers, including certain orchids, such as *Platanthera bifolia* in Norway, whose floral structure and fragrance is designed to attract moths. The Pine Hawk Moth also pollinates *P. chlorantha*, and research has shown that the release of that orchid's floral volatiles (chemicals that attract pollinators) coincides with the moth's flight time.

FEEDING ON CONIFERS

Females disperse their eggs widely, laying two to three at a time on pine needles and twigs, where the neonates hatch and are initially yellow with a black-and-yellow head. Early instars feed on a needle's surface, but later instars, which are cryptically colored with striated green patterning, consume all parts of the needle. Mature larvae develop a brown stripe down the middle of their back and, prior to pupation, become brown or gray, then crawl down the tree trunk and bury themselves to pupate. While there is one generation annually in the north of its range, two can occur farther south. Many of the other 27 species in the widespread genus *Sphinx* are also associated with coniferous plants.

→ The Pine Hawk Moth is only one of several pine-feeding hawk moths. Others have distinctive geographic and ecological preferences, and include the Chinese Pine Hawk Moth (*Sphinx caligineus*) and the Asian Pine Hawk Moth (*S. morio*). Caterpillars of the similar-looking Doll's Sphinx (*S. dollii*) and Sequoia Sphinx (*S. sequoiae*) develop on junipers and cedar in the southwestern United States.

BUPALUS PINIARIA

The Bordered White

Feathery antennae

SCIENTIFIC NAME	*Bupalus piniaria* (Linnaeus, 1758)
FAMILY	Geometridae
NOTABLE FEATURES	Slender body, feathery antennae
WINGSPAN	1³⁄₁₆–1⁹⁄₁₆ in (30–40 mm)
SIMILAR SPECIES	*Bupalus vestalis*, but paler and occurs in China

Common from western Asia and Europe to North Africa, and eastward into Siberia, the Bordered White moth (or Pine Looper) flies in a single annual generation in the spring and lives for up to two weeks. A single female can lay around 180 eggs in small batches, arranged along a pine needle in strings of several eggs, which hatch three weeks later. The caterpillars develop through five instars, concealed against the cones by their cryptic green coloring interspersed with pale lines, before finally pupating in leaf litter. The larvae attack a variety of conifers, including pines, spruce, fir, and larch.

COLOR VARIATIONS

The adult male has a broad, brown border along the forewing and more feathery antennae than the female, whose coloration is also more subtle. The moth's forewings are yellow to brown, while the hind wings are orange. The white streak on the underside gives the moth a characteristic, butterfly-like appearance, enhanced by its day-flying habit and its position when at rest, with wings closed above the thorax. Its borders and spots may differ according to geographic location, and its background color can vary from white in the north to deep yellow in the south of its range.

PEST OR FOOD FOR BIRDS?

While all stages of these moths are important food for various vertebrate predators, especially nesting birds, the larvae are considered a pest on pine in some areas. White Bordered moth numbers can peak every five to seven years as part of the species' natural cycle. Historically, control methods have had disastrous results; 50 years after an area of UK pine woodland was aerially sprayed with DDT to eradicate *Bupalus piniaria* caterpillars, residues remain in the woodland ecosystem, causing a significant decline in wildlife.

→ The male Bordered White moth has large, feathery antennae that help it locate females.

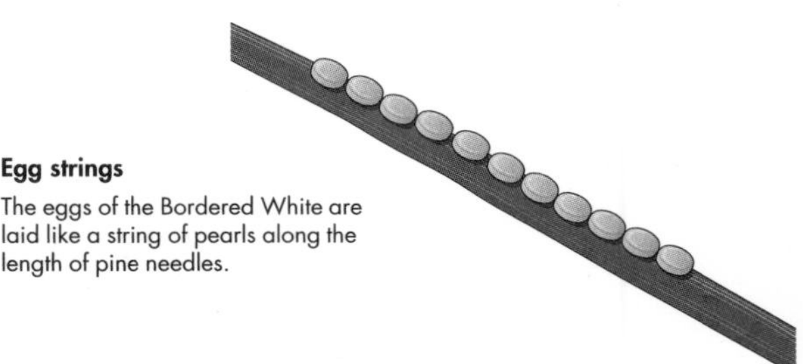

Egg strings

The eggs of the Bordered White are laid like a string of pearls along the length of pine needles.

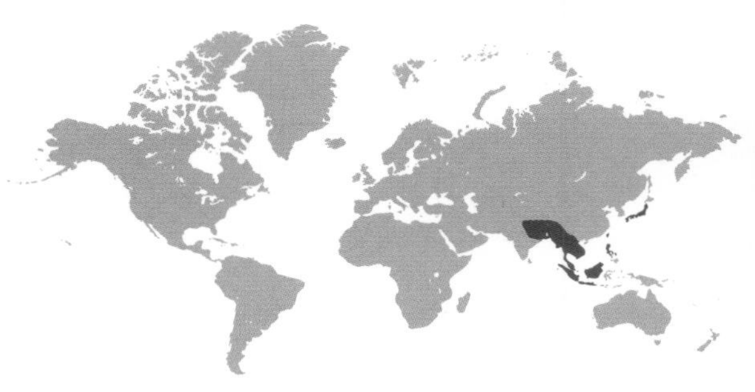

MILIONIA BASALIS

Milionia basalis

Striking Japanese geometrid

SCIENTIFIC NAME	*Milionia basalis* (Walker, 1854)
FAMILY	Geometridae
NOTABLE FEATURES	Dark with iridescent, geometric, electric-blue patterning; black-specked bands of red, orange, or yellow across wings
WINGSPAN	2 – 2$^{3}/_{16}$ in (50–56 mm)
SIMILAR SPECIES	*Milionia fulgida* (Borneo and Java) and several subspecies, such as *M. b. pyrozona* (Malaysia, Myanmar) and *M.b. pryeri* (Japan)

Found in Asia from the northeastern Himalayas to Japan, this day-flying moth is one of the most beautiful representatives of the family Geometridae, which largely includes cryptic species.

EXPANDING SPECIES

The moth has been increasing its range: in Japan, for example, it was once found only in the subtropical evergreen forests of the southern Nansei Islands, but has now spread northward to Kyushu, where up to four generations are produced each year. The moths, which feed on nectar from flowers of the powder-puff tree (*Barringtonia racemosa*) in Japan and those of the yellow tea tree (*Leptospermum flavescens*) in the mountains of Peninsular Malaysia, among others, peak in spring and mid-summer but, in the mild climate of Kyushu, emerge even in early winter.

APOSEMATIC COLORATION

Females lay eggs on the leaves of their host plants—*Dacrydium* and *Podocarpus*; their larvae can be a pest on *Podocarpus*, which is frequently cultivated for use in construction. When mature, the caterpillars drop down on a thread from the tree and pupate in soil beneath the host plant.

The larvae, which have an orange head and orange final segments, are chemically defended, acquiring their toxic properties (passed onto the adults) from *Podocarpus* leaves. They use this aposematic signaling of their toxicity as a defense against predators, rather than camouflage, which is common in other geometrids. The chemicals that the caterpillars obtain from *Podocarpus* are quite potent: in lab research, they killed predatory stink bugs that were fed *M. basalis* larvae.

→ *Milionia basalis*, the beautiful geometrid that feeds on *Podocarpus* as a caterpillar, is "puddling" here—absorbing microelements from the wet soil—behavior frequently observed in males of diurnal Lepidoptera.

CYDIA PIPERANA

Ponderosa Pine Seedworm Moth

Shiny but cryptic

SCIENTIFIC NAME	*Cydia piperana* (Kearfott, 1907)
FAMILY	Tortricidae
NOTABLE FEATURES	Shiny, but cryptically colored, with transverse stripes on the hind wings
WINGSPAN	⅝–¾ in (16–21 mm)
SIMILAR SPECIES	Other *Cydia spp.* especially Longleaf Pine Seedworm Moth (*C. ingens*), Eastern Pine Seedworm Moth (*C. toreuta*), and *C. erotella*

Although a member of the Tortricidae family of 11,000 species known as leaf roller moths, larvae of the Ponderosa Pine Seedworm Moth do not live within rolled leaves; the caterpillars mine pine cones instead.

PINE LOVERS

The species' common name reflects its larval diet of seeds of the ponderosa pine (*Pinus ponderosa*), and they also feed on seeds of the Jeffrey pine (*P. jeffreyi*). The brownish-gray moths, whose hind wing stripes help to camouflage them against pine bark, are found in North America west of the Rocky Mountains. The larvae may overwinter in the pith of a cone, and the adults fly between February and June, depending on their location.

TREATED AS PESTS

The moth is considered a pest, and therefore synthetic attractant chemicals—(E)-9- and (Z)-9-dodecenyl acetates—have been successfully developed to trap them. An almost identical chemical was identified as a major component of the sex pheromone of the Eastern Pine Seedworm (*Cydia toreuta*), a similar species that feeds on red pine (*Pinus resinosa*) and jack pine (*P. banksiana*) in the Eastern United States and Canada.

A survey of ponderosa pine cones in northern Arizona showed that the *Cydia piperana* caterpillar destroys only 1.3–7.6 seeds per cone as it develops, but in British Columbia the caterpillar may destroy up to 50 percent of the seeds.

→ The Ponderosa Pine Seedworm Moth is one of over 200 species in its genus, the most infamous of which is the Coddling Moth (*Cydia pomonella*) that feeds on fruit such as apples. In Europe, the Ponderosa Pine Seedworm Moth feeds on damaged pine bark wherever fungus takes hold.

MOTHS IN AGROECOSYSTEMS & AROUND HOMES

Moth survival in a human world

A thousand years ago, only 4 percent of the world's habitable land was farmed; the figure today is 50 percent. Vast tracts of land that were once natural biomes are now cultivated: prairies have been planted with wheat, corn grows where there was once forest, soybean and palm plantations have replaced jungles. Humans have transformed Earth's natural landscape to sustain ever-increasing populations, so it is little surprise that other creatures, including moths, continue to make their own bid for survival.

CHANGING THE HOSTS

The crops now farmed intensively are modified varieties of wild plants that Lepidoptera have fed on for millennia, as are the highly prized flowers, shrubs, and palms exported to gardens and front yards across the world. The Diamondback Moth (*Plutella xylostella*) and the Cabbage Looper (*Trichoplusia ni*) that once sought out wild cruciferous plants have continued to feed on cultivated crucifers whether they are collard greens or cabbageheads, because what attracts them are the secondary chemicals that give the vegetables their unique, slightly bitter taste. The Tobacco Hornworm (*Manduca sexta*) remains as attracted to nicotine as it was when the tobacco plants grew wild. The olive tree, *Olea europaea*, cultivated for several thousand years, had a wild ancestor (*O. europaea silvestris*) that attracted species including the Olive Moth (*Prays oleae*), whose larvae still mine olive tree leaves.

As a proportion of all Lepidoptera species, those that target agricultural crops are relatively few in number, but they and many other so-called insect "pests" now cost world economies some $70 billion annually in lost crop production, while millions more are spent researching ever more sophisticated strategies for limiting their attacks.

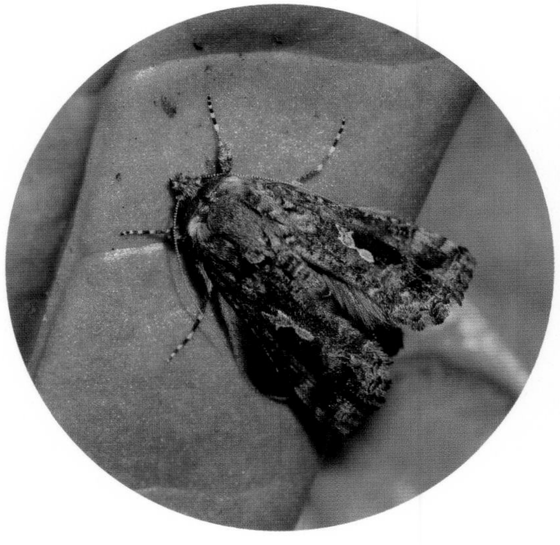

↑ The Cabbage Looper (*Trichoplusia ni*) adult can lay numerous eggs, usually singly, and its ferocious larvae can feed on a large variety of hosts, cruciferous plants such as cabbage being just a small subsample.

→ The Tobacco Hornworm (*Manduca sexta*) caterpillar is one of the best studied, thanks to its large size. When mature, the caterpillar wanders around searching for a suitable place to pupate underground, then forms a pupal cell reinforced with caterpillar saliva (these larvae don't spin silk).

War on "pest" moths

In the battle against highly invasive species such as the Corn Earworm (*Helicoverpa zea*), which favors corn (see also pages 278–279), or the ubiquitous Codling Moth (*Cydia pomonella*), whose larvae feed on orchard fruits, entomologists and growers practice integrated pest management (IPM).

RESISTANCE

Pesticides have been widely used but have major drawbacks. Caterpillars adapt quickly to the chemicals because, if only a resistant type survives, the next generation of larvae is much more likely to be resistant. Thus, pesticides can become impotent within just a few years after introduction (in much the same way that germs develop the ability to resist antibiotics). Many insecticides also kill off caterpillars' natural enemies, so the pests proliferate faster once they gain resistance to chemicals.

MIGRANT PESTS

Insects also tend to become pests when they are transplanted from their native distribution areas to new areas that lack the natural enemies that would normally keep them in check; this is why strict quarantine rules are imposed at international airports, despite which pests sneak in, frequently with imported produce or ornamental plants. Researchers, working with agricultural crop producers, sometimes import a pest's native enemies, and breed and release them in a bid to suppress an emerging pest species. Among the best-known natural enemies of caterpillars are parasitic wasps that tend to attack a specific moth genus or even a single species. The danger of introducing such caterpillar enemies, however, is that they might also target native fauna, creating new problems.

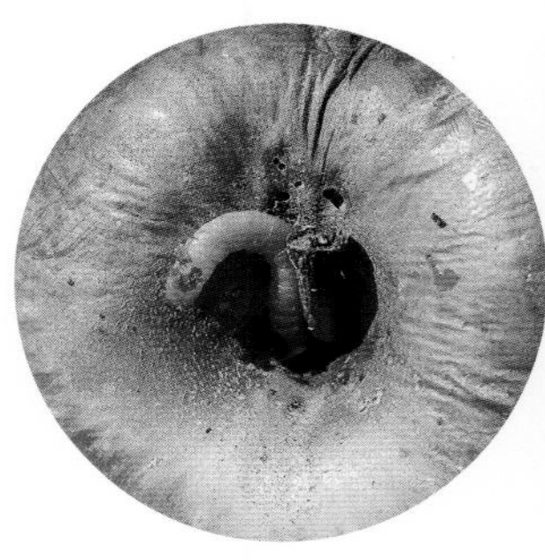

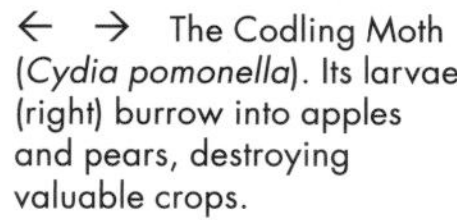

← → The Codling Moth (*Cydia pomonella*). Its larvae (right) burrow into apples and pears, destroying valuable crops.

INGENIOUS DEFENSES

The Subflexus Straw Moth (*Chloridea subflexa*), which is practically indistinguishable from its infamous Tobacco Budworm relative (*C. virescens*), feeds exclusively on ground cherries (*Physalis* spp.), after making an entrance hole in the cover leaves surrounding the fruit. These then provide a shelter for the caterpillar. Scientists have shown that while most plants give off a chemical warning signal when caterpillars feed on them, by feeding on these fruits, the Subflexus Straw Moth caterpillar can remain "invisible" to parasitoid wasps that hunt by scent, developing chemical crypsis thanks to its unique diet. While the Straw Moth is relatively rare, its infamous Tobacco Budworm sister species, which damages anything from tobacco, cabbage, and cantaloupe to clovers, cotton, and peas, is widespread. It is crucial to understand both biology and taxonomy of pests to be able to distinguish between the "good" and the less so (from the human point of view) species of moths.

Feeding on physalis

Chloridea subflexa caterpillars have developed the unique ability to live on physalis fruits which lack the linolenic acid that larvae of other moth species require in their diet. In the absence of this chemical, parasitoids, which would normally attack the larvae, cannot protect them.

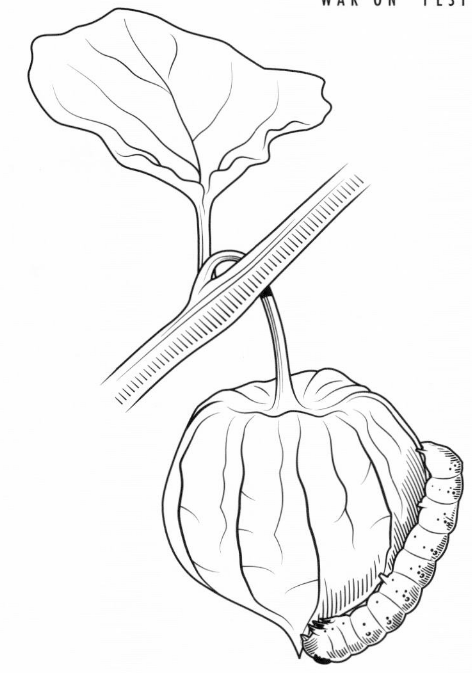

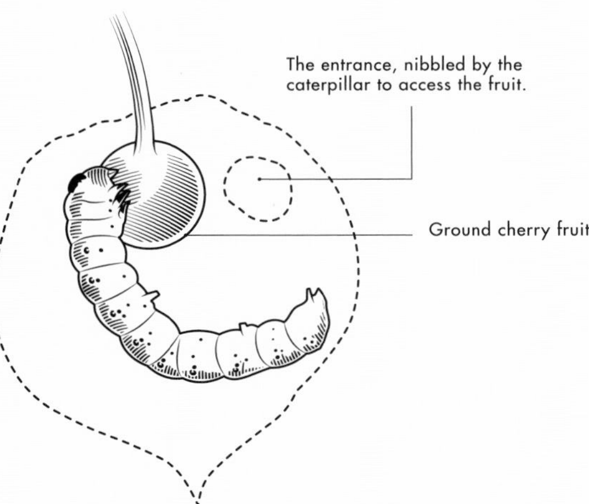
The entrance, nibbled by the caterpillar to access the fruit.
Ground cherry fruit

Before pupating, the caterpillar may reinforce the papery physalis leaves with silk to keep the pupa secure.

Food and shelter

Chloridea subflexa caterpillars have developed the unique ability to live on fruits of the ground cherry (*Physalis* spp.). Their chemical crypsis protects them from parasitoids because the plants they feed on produce the SOS signal of volatile chemicals that parasitoids detect only when their leaves, rather than their fruits, are attacked. In addition, the lantern-like physalis leaves that protect the fruit provide shelter for the caterpillar as it develops and pupates.

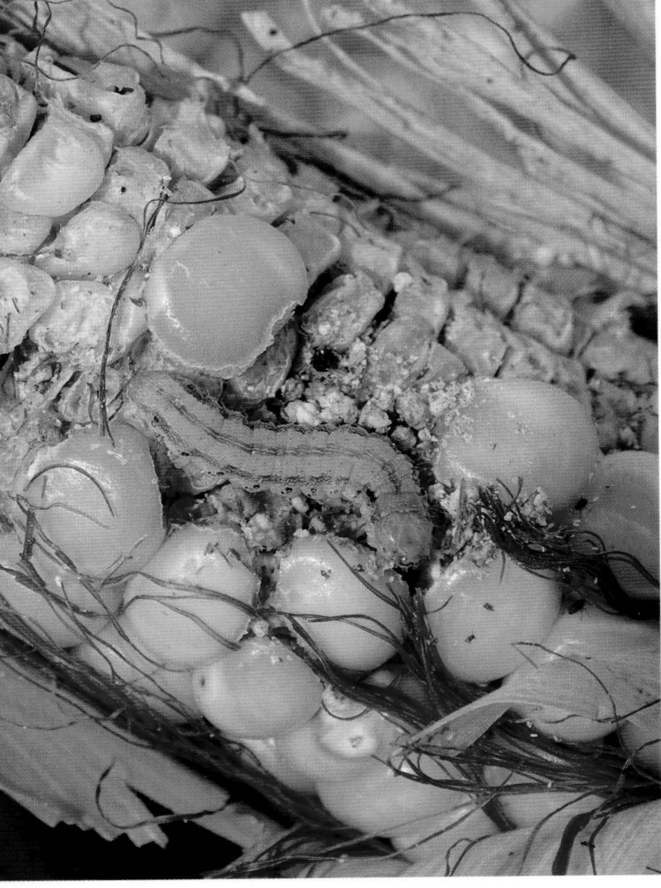

PHEROMONE TRAPS

One major element of wise pest control is monitoring: when to spray pesticides is of key importance to gain maximum benefits with the fewest adverse side effects. Pheromone traps have become part of such practice, so researchers are frequently engaged in collecting pheromones from moths, analyzing their chemical composition, determining their effectiveness, and finally synthesizing them for commercial use. Pheromone traps can reveal how many pests are present in a field or orchard. While trapping all the pests present is problematic (as almost always it is males that are caught, so females remain free to lay their eggs), dispensing synthetic pheromones in an area can sometimes disrupt mating and thus reduce infestations.

NOVEL TECHNIQUES IN PEST COMBAT

In some cases, scientists have employed microorganisms, such as bacteria and viruses that attack insects, replicating them in laboratories and then using them as "natural" pesticides. For instance, the granulosis virus, a member of the baculovirus group of viruses, has been used against the Indian Meal Moth (*Plodia interpunctella*) that feeds on grains and a variety of processed products. *Bacillus thuringiensis* (Bt), a bacterium that is lethal to many immature insects, including moth caterpillars and pupae,

↖ A pheromone trap used in an orchard to attract and kill plum moths or Codling Moths (*Cydia pomonella*). This is an environmentally safer way of reducing infestations than pesticides.

← The Corn Earworm (*Helicoverpa zea*), found from Canada to Argentina, is one of the costliest New World pests. Thanks to its ability to feed on a variety of crops and its resistance to many pesticides, additional IPM (integrated pest management) strategies have to be employed, such as biological control using parasitoids and deep plowing of soil to kill the pupae.

↗ The efficacy of baculovirus as a control method is demonstrated here, as caterpillars of the Codling Moth turn black and die after infection by the virus.

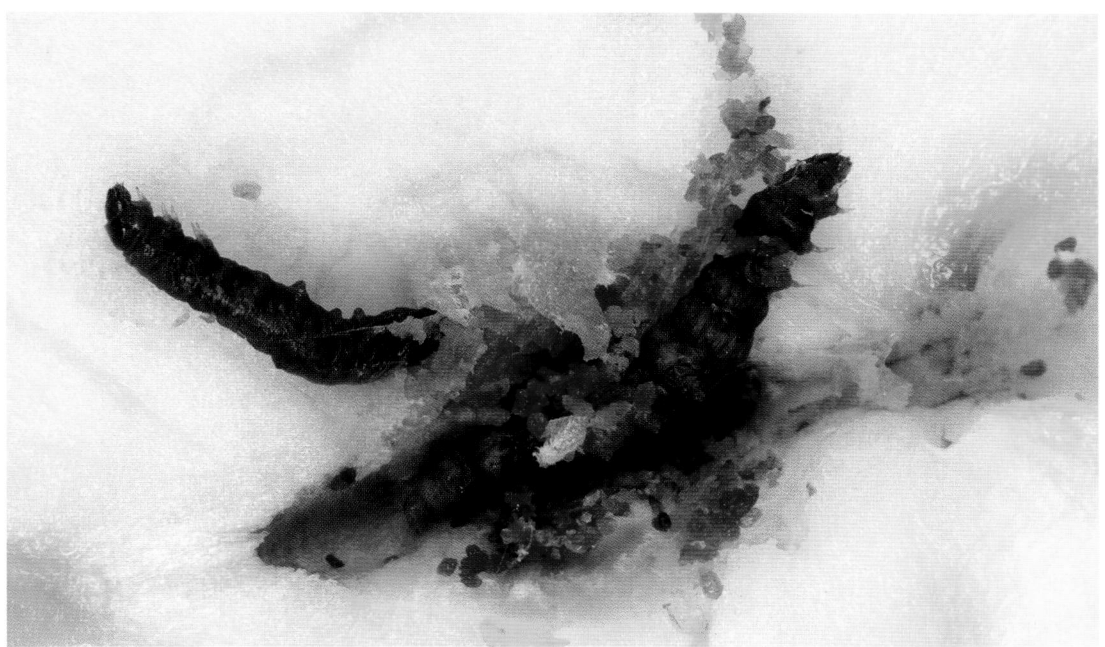

has also become an effective tool for controlling many pest species. However, as with pesticides, insects can develop resistance to the pathogens: for example, exposing populations of mealworm moths to low levels of the granulosis virus noticeably increased resistance in their offspring.

A further strategy is to release sterile moths in the hope of reducing populations. In the past, this was done by breeding and then irradiating the insects before their release. In 2020, the Nobel Prize in chemistry went to researchers who discovered CRISPR (Clustered Regularly Interspaced Short Palindromic Repeats)—genome editing technology, which has a wide variety of medical research applications. It is already used to silence genes in Lepidoptera in a bid to understand their function, and it has been proposed that targeted genome editing could be useful in the future to combat pests. For instance, mass producing individuals able to mate with wild counterparts but unable to reproduce, would lead to infertile eggs and a reduction of infestations.

An alternative approach has been to transform the genetics of crops by selecting for the more crop-resistant varieties—a practice that farmers have employed, intentionally or accidentally, for thousands of years, when sowing seeds from plants that have survived pest attacks. Now, there are additional shortcuts in the process, which even include incorporating bacterial DNA into the DNA of crop species, so that they can produce their own insecticides.

Despite all the negative effects of the few pests that tarnish the image of the more than 100,000 beneficial moth species, much of what we know about moth biology today is due to the researchers engaged in studying these "economically important" moths whose feeding habits are so disruptive to agriculture. Understanding that the word "pest" is a human construct, that no creature is a pest in the absence of human activity, and that humans create perfect environments for some species to become pests, are among the first steps toward appreciating these remarkable insects.

Moths in storage areas and homes

Some moths, such as the Indian Meal Moth (*Plodia interpunctella*), have evolved to feed on processed grain products such as flour or breakfast cereals, while the notorious Common Clothes Moth (*Tineola bisselliella*) is notable for its ability to digest keratin—a nutritious protein that is present in natural fibers.

↑ Indian Meal Moths (*Plodia interpunctella*) feed on cereal crops but are frequently found in houses and storage facilities, too, where they infest plant-based dry foods, for which they are also known as pantry moths, flour moths, or grain moths.

GRAIN MOTHS

Where grains and grain products are stored, insects are swift to follow, including several moth species. The gelechiid Angoumois Grain Moth (*Sitotroga cerealella*) was originally described in France, but its true origins are unknown. Today this species is widespread throughout the world. Tiny, maggot-like larvae burrow into the germ of growing and stored cereal grains, which then become unusable. The female is able to lay close to 100 eggs, positioning them between grains, so the species is easily spread during transportation. Up to three larvae may develop and feed within a single grain (although one is more usual), and sometimes create silken tunnels to nearby food sources. The curved hind wings of the brownish-gray adults are an unusual and distinguishing feature.

The little, whitish caterpillar of the Indian Meal Moth (*Plodia interpunctella*), distributed on every continent except Antarctica, also feeds on grains, as well as cereals, nuts, and flour, and can chew through plastic or cardboard. The non-feeding adults, which have brownish forewings, white hind wings, and a brown-white body, reproduce on both foods and clothing.

In a remarkable twist of fate, the grain moths became not only pests but also servants of humans: many beneficial species are reared on the eggs of the Angoumois Grain Moth, from common green

lacewings to *Trichogramma ostriniae* (the tiny egg parasitoid wasps that are used to control another terrible pest, the European Corn Borer (*Ostrinia nubilalis*). In the bid to understand the genetics of Lepidoptera for the sake of future successful IPM practices, the Indian Meal Moth has become a model research organism, with its complete genome sequenced and almost 85,000 unique genes identified.

CLOTHES MOTHS

Of the various moths whose caterpillars feed in human homes, the Common Clothes Moth (*Tineola bisselliella*) and the Casebearing Clothes Moth (*T. pellionella*)—both in the family Tineidae—are among the best known and most widespread. Their evolutionary success has been greatly assisted by humans— the spread of the Common Clothes Moth has been documented on numerous occasions, as for example in *Notes by a Naturalist on the "Challenger"* (1879), Henry Nottidge Moseley's account of his findings during a British scientific voyage. As far back as the fifth century BCE, the ancient Greek playwright Aristophanes wrote of the Clothes Moth's destructive capabilities as did Pliny in Rome at the dawn of Christianity. While the Common Clothes Moth seems to have established itself as far north as the tundra regions, it appears to have had much less success colonizing the tropics.

The Common Clothes Moth's larvae have a varied diet and have been found not only on wool, but also in fishmeal, dried meats, drugs containing albumin, insect remains, and even on mummified human corpses. Their evolutionary success is largely attributed to their ability to digest keratin. The small beige or buff-colored adults have narrow forewings and can be hard to distinguish from other tineids without specialized knowledge and the use of dissection or molecular techniques. Females can lay as many as 200 eggs, and the larvae conceal themselves in silken tunnels, developing over five to six instars, and on occasions going through many extra molts. Like all moths, their final size greatly depends on the quality of food, with pupal weight ranging from 3 to 10 mg and wingspan from ⅜ to ⅝ in (9–16 mm). They prefer soiled, rather than clean fabrics, and may be deterred by lavender spray or naphthalene moth balls or flakes.

↖ The Angoumois Grain Moth (*Sitotroga cerealella*) develops within a single grain kernel, hollowing it out as shown here.

↑ The Common Clothes Moth (*Tineola bisselliella*) is widely known for the damage its larvae causes to clothes, especially those made from wool.

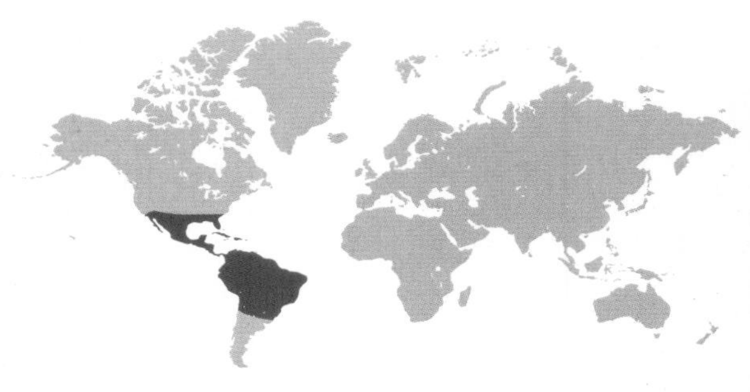

TERASTIA METICULOSALIS

Erythrina Borer

Coral bean slayer

SCIENTIFIC NAME	*Terastia meticulosalis* (Guenée, 1854)
FAMILY	Crambidae
NOTABLE FEATURES	Resting posture that resembles a mantis head
WINGSPAN	1–1⅘ in (25–46 mm)
SIMILAR SPECIES	*Terastia africana* (Africa), *T. subjectalis* (Asia)

Named for the plants its larvae feed on, the Erythrina Borer is widespread from the Carolinas to California in the United States, to Argentina in South America. The adults vary greatly in size as a result of their diet at earlier life stages; in Florida, those that develop in spring on the nutritious beans of *Erythrina herbaceae* are larger than those whose larvae feed inside the stems in the summer and fall. The moths, which live for several weeks, are strong fliers and can disperse long distances.

DISTINCTIVE POSTURE

The moths' posture at rest is distinctive: as they point their abdomen upward, the abdomen scales form hollow chambers, creating an impression of a second head which, viewed from a certain angle, resembles that of a praying mantis.

ERYTHRINA HOST PLANT

Females lay their eggs on the fresh shoots or forming pods of their host plants, including all coral trees or coral beans (*Erythrina* spp.), highly prized by landscapers for their flame-red flowers. Newly hatched *Terastia meticulosalis* larvae bore into pods or stems and hollow them out as they develop, leaving them only to pupate, when they spin a tough double cocoon near the host plant or among its leaves.

AN EMERGENT PEST

In general, Erythrina Borers are in balance with their environment, though historically their prevalence has limited coral tree cultivation in Florida. Once rare in California, where the trees are popular ornamentals, the moths have become common in the last decade, attacking young plants in nurseries. As spraying does not affect larvae feeding inside stems, pheromone traps could become a potential control method in the future.

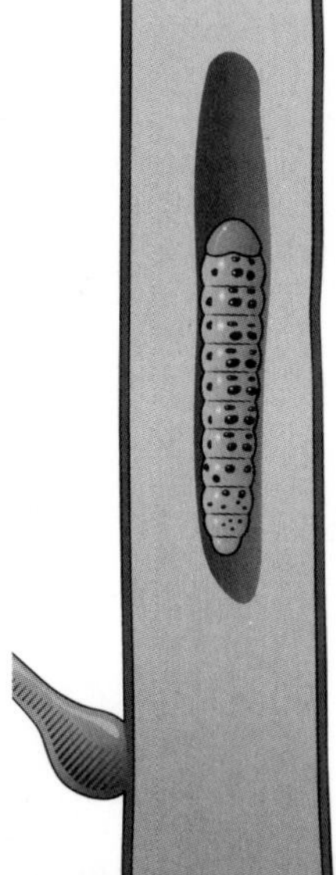

Stem borer

The Erythrina Borer caterpillar burrows inside a leaf petiole first, and then makes its way into a stem and hollows out its center, causing the top of the young plant or fresh shoots of a mature tree to die off.

→ An Erythrina Borer moth in Florida. When at rest, this moth's abdomen is pointed upward, and swollen as a result of special scales. The posture and hollow protrusions may have evolved to mimic the head and eyes of a predator such as a praying mantis.

HYPHANTRIA CUNEA

Fall Webworm

Silken web weavers

SCIENTIFIC NAME	*Hyphantria cunea* (Drury, 1773)
FAMILY	Erebidae
NOTABLE FEATURES	Snow-white moth, sometimes with small, dark spots
WINGSPAN	1⅓ – 1⅜ in (35–42 mm)
SIMILAR SPECIES	Other white erebids, such as *Spilosoma* and *Euproctis* spp.

Widely studied because of its invasive larvae, the Fall Webworm moth is native to North America but present in Central America and, since the 1940s, much of Europe, as well as eastern Asia, from Mongolia to Japan. Wherever the species has been introduced (often during transportation of its food plants), it has then spread by natural means, steadily increasing its distribution area.

NOCTURNAL HABITS

The moths tend to emerge in the evening, and mating occurs before sunrise. Females lay up to 1,900 eggs on various deciduous trees in batches of differing sizes, covering them with hairs from their abdomen. While populations may produce only one generation annually in northern parts of their range, farther south there may be up to four. The adults do not feed and live for around a week.

FALL WEBS

Fall Webworm caterpillars are hairy and initially yellow with black dots, but when mature have subtle darker colors, with red heads in some broods. In the fall, they spin formidable silken nests, extending them from the tips to envelop whole branches, then move in a group to a new branch when the previous one is defoliated, sometimes covering an entire tree with their webs. The density of the nests helps trap heat to maintain a constant temperature as the larvae develop and protects them against predators. When mature, if one senses danger, all may wiggle in synchrony creating the surreal impression of dancing caterpillars. To control them, in addition to pesticides, nests are removed and destroyed as soon as they appear.

→ Fall Webworm moths may be pure white, or speckled to a varying degree.

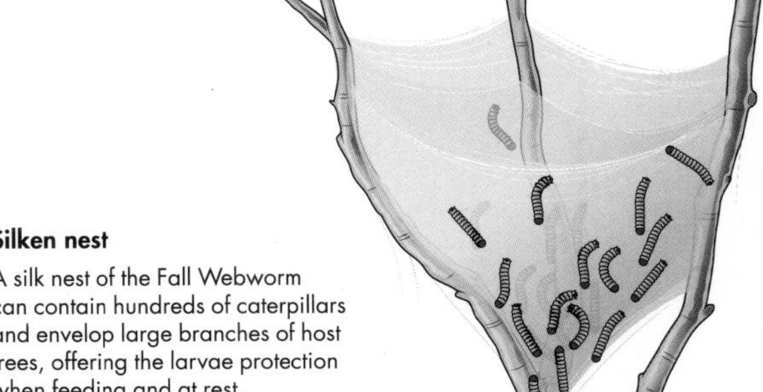

Silken nest

A silk nest of the Fall Webworm can contain hundreds of caterpillars and envelop large branches of host trees, offering the larvae protection when feeding and at rest.

ZEUZERA PYRINA

Wood Leopard Moth

Spotted, white pest

SCIENTIFIC NAME	*Zeuzera pyrina* (Linnaeus, 1761)
FAMILY	Cossidae
NOTABLE FEATURES	White with iridescent dark spots; antennae first serrate, but tapered at tip
WINGSPAN	1⅓–2⅜ in (35–60 mm)
SIMILAR SPECIES	Five other *Zeuzera* spp.; in the US, erebid Giant Leopard Moth (*Hypercompe scribonia*)

Known largely for its larvae, considered pests of orchard and olive trees, this large, rather lovely moth is found throughout Eurasia and North Africa, and was first recorded in the United States in the nineteenth century. Both sexes are similar, but the male is slightly smaller and has broader, more feathery antennae. They mate soon after emergence, and the female may lay several hundred eggs, often deposited in clusters in the crevices of tree bark.

WOOD-EATING CATERPILLARS

The young larvae soon disperse and begin to tunnel through the living wood, boring into new growth—especially young shoots and tips of branches. They then feed on the wood for two to three years, sometimes killing a tree before they reach pupation; in captivity, however, they have been raised within three or four months on a diet of soybean, milk powder, and yeast, mixed with sugar-beet pulp.

CONTROL METHODS

Growers find it difficult to combat carpenter moths and other wood-boring species as pesticides cannot reach them within their tunnels. In Bulgaria, where the larvae were destroying up to 30 percent of trees in older apple orchards, many have been caught in pheromone traps between June and September, indicating the best time for using pesticides to reduce adult populations. In Iran, where they are a pest of walnut trees, researchers have explored the use of nematodes (worms) that attack insects as a possible control method, while in Italian apple orchards, dispensing synthetic moth pheromones has proved effective for disrupting mating, resulting in lower levels of infestation.

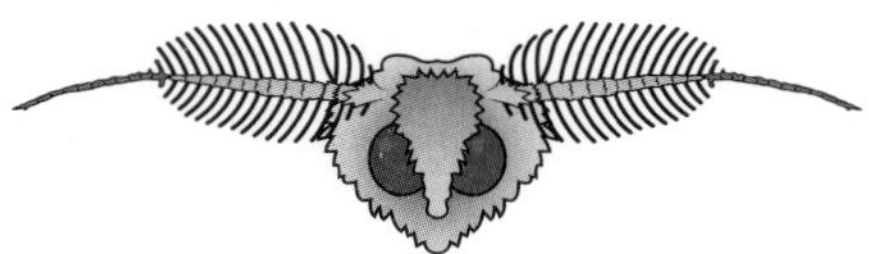

Tapering antennae

Antennae of the Wood Leopard Moth, especially those of the male, are wide at the base and highly serrate, but taper toward their tips.

→ *Zeuzera pyrina*, as larvae, are wood-boring and can damage a wide range of fruit trees, from olive and walnut trees in the Middle East to apple orchards in Italy. Naturally, the moth can develop in over 100 different species of wild host plants too, such as maples and elms, but do not cause serious damage to tree populations.

HELICOVERPA ZEA

Corn Earworm

Invasive crop destroyer

SCIENTIFIC NAME	*Helicoverpa zea* (Boddie, 1850)
FAMILY	Noctuidae
NOTABLE FEATURES	Often a central dark spot on forewings and hind wings
WINGSPAN	1¼ – 1¾ in (32–45 mm)
SIMILAR SPECIES	Other *Chloridea* spp., such as the Tobacco Budworm (*C. virescens*).

The undistinguished, yellowish-brown Corn Earworm adult produces rapacious larvae—among the most destructive insects in the Americas, consuming a variety of wild plants and crops, including corn, soybeans, tomatoes, and cotton.

LARGE POPULATIONS

The mainly nocturnal moth hides by day in vegetation, venturing out only to feed on nectar. Its usual lifespan is 5 to 15 days, but occasionally a month. From the third day after eclosing until their death, females lay between 500 and 3,000 eggs on ears of corn and many other wild and cultivated species. The moth, which flies fast, populates northern parts of its range every year via migration and, carried by wind currents, can reach high elevations. The Corn Earworm has difficulty surviving harsh winters and tends to reproduce and overwinter in warmer climates.

CANNIBALISTIC LARVAE

As *Helicoverpa zea*'s common name suggests, the larvae feed on maturing corn ears, as well as the leaves and flowering heads of corn and other crops. The young caterpillars feed in clusters at first, then those that grow faster eat their siblings, leaving only several per site to mature, and will even attack and eat larvae of other Lepidoptera.

HIGH COST OF CONTROL

Crop losses and pest management, exacerbated by the larvae's fast-developing resistance to chemicals, cost up to $250 million dollars a year in the United States alone. In addition to pesticides, *H. zea* populations can be suppressed by parasitoid wasps such as *Microplitis*, and by deep plowing of fields between planting, crop rotation, and the use of crops genetically modified for pest resistance.

→ A migratory moth that overwinters in warmer realms, even in deserts, the Corn Earworm relies on nectar from flowering plants such as thistles (here) for energy to fuel its long-distance flight.

GLOSSARY

aposematic Defensive bright coloration, signaling distastefulness.

Batesian mimicry When a harmless, edible, usually rarer species (such as a clearwing moth) resembles a distasteful, often more common species (such as a wasp), thus gaining some protection from predators.

biome A set of similar habitats.

bursa copulatrix A sac in the female moth's abdomen where a male's spermatophore (mass of sperm) is stored after mating.

chitin Fibrous material made of polysaccharides (carbohydrates) that makes up the exoskeleton (outer layer) of arthropods (insects, spiders, or crustaceans).

color dimorphism Differences in color; color differences between a male and a female are known as "sexual dimorphism."

cremaster Hooks (sometimes spines) on the last abdominal segment of a pupa, often used to attach the pupa to a silk pad or a cocoon.

crepuscular Active at dusk.

crochets Hooks on the prolegs of a caterpillar enabling it to grip the substrate (surface on which it lives).

cryptic Colored to blend in with the substrate, as opposed to aposematic.

cuticle Epidermis or skin of an insect.

diapause A state of suspended development and prolonged inactivity, induced by environmental changes and produced by a major shift in hormonal activity (winter diapause, for example).

diurnal Active during the day.

eclose To emerge from a pupa (noun: eclosion).

estivate To avoid heat of the summer by entering a period of dormancy in a cooler place, such as a cave or at a high elevation.

exoskeleton The cuticle of an insect that protects and supports its body, and to which its internal muscles attach.

frass Insect droppings.

hemolymph Insect equivalent of blood.

instar The stage of development between molts in a caterpillar.

lek/lekking Gathering of male insects, for example, to display and compete for the attention of females.

maxillary palpi Sensory appendages of the mouthparts used to "smell" food and assist in eating.

morphology Structures of organisms; study of these structures and how they relate to each other in different organisms (comparative morphology) or how they function (functional morphology).

Müllerian mimicry When two or more species, all distasteful to predators, resemble each other to indicate that they are defended.

New World North, Central, and South America.

Old World Africa, Europe, and Asia, or Afro-Eurasia.

outbreaks For insects, a periodic sharp increase in numbers that lasts a few generations.

oviposition/ovipositor Laying eggs/ organ for laying eggs.

parasitoid A parasite (wasp or fly, for example) whose larvae feed on a host (such as a caterpillar) developing inside it and eventually killing it.

polyphagous Capable of eating many different types of food (different, unrelated plant species, in the case of moth caterpillars).

proboscis Tubelike mouthpart of Lepidoptera, adapted for sucking fluids.

prolegs Fleshy projections of the body on the abdominal segments of caterpillars or similar larvae that enable them to hold on to a substrate.

sclerotized Hardened (usually describing a cuticle).

sequester To accumulate by selectively storing (used in the context of protective chemical compounds that some caterpillars accumulate from their food).

sensilla Microscopic organs of arthropods, used to sense the world (taste, smell, for example).

spinneret In spiders and other insects, an organ for spinning silk which is secreted as liquid by silk glands and usually solidifies to become a silk thread; in caterpillars, it is a mouthpart.

serrate Comb-shaped (describes moth antennae).

setae Individual hairs or bristles, usually associated with sensing.

stemma (pl. stemmata) An insect's simple eye.

stridulation Of an insect, a sound usually made by rubbing rough surfaces of legs, wings, or other body parts together.

substrate Surface on which an organism lives.

tribe A taxonomic category below subfamily but above genus level.

type specimen A specimen, or series of specimens, on which the description of a new species is based. Includes a holotype—the single specimen designated as a "standard" for a particular species—while paratypes are a series of specimens from which a neotype (new holotype) can be selected if the original holotype is lost.

MOTH FAMILIES

A list of the major moth families, with the number of described genera and species. The families mentioned in this book are highlighted in bold.

Acrolophidae (Busck, 1912) 5 genera, 300 species

Adelidae (Bruand, 1850) 5 genera, 294 species

Alucitidae (Leach, 1815) 9 genera, 216 species

Anthelidae (Turner, 1904) 9 genera, 94 species

Apatelodidae (Neumoegen & Dyar, 1894) 10 genera, 145 species

Argyresthiidae (Bruand, 1850) 1 genus, 157 species

Autostichidae (Le Marchand, 1947) 72 genera, 638 species

Batrachedridae (Heinemann & Wocke, 1876) 10 genera, 99 species

Blastobasidae (Meyrick, 1894) 24 genera, 377 species

Bombycidae (Latreille, 1802) 26 genera, 185 species

Brachodidae (Agenjo, 1966) 14 genera, 137 species

Brahmaeidae (Swinhoe, 1892) 7 genera, 65 species

Bucculatricidae (Fracker, 1915) 4 genera, 297 species

Callidulidae (Moore, 1877) 7 genera, 49 species

Carposinidae (Walsingham, 1897) 19 genera, 283 species

Castniidae (Boisduval, 1828) 34 genera, 113 species

Choreutidae (Stainton, 1858) 18 genera, 406 species

Coleophoridae (Bruand, 1850) 5 genera, 1,386 species

Cosmopterigidae (Heinemann & Wocke, 1876) 135 genera, 1,792 species

Cossidae (Leach, 1815) 151 genera, 971 species

Crambidae (Latreille, 1810) 1,020 genera, 9,655 species

Douglasiidae (Heinemann & Wocke, 1876) 2 genera, 29 species

Drepanidae (Boisduval, 1828) 122 genera, 660 species

Elachistidae (Bruand, 1850) 161 genera, 3,201 species

Endromidae (Boisduval, 1828) 12 genera, 59 species

Epermeniidae (Spuler, 1910) 10 genera, 126 species

Erebidae (Leach, 1815) 1,760 genera, 24,569 species

Eriocottidae (Spuler, 1898) 6 genera, 80 species

Eupterotidae (Swinhoe, 1892) 53 genera, 339 species

Euteliidae (Grote, 1882) 29 genera, 520 species

Gelechiidae (Stainton, 1854) 500 genera, 4,700 species

Geometridae (Leach, 1815) 2,002 genera, 23,002 species

Glyphipterigidae (Stainton, 1854) 28 genera, 535 species

Gracillariidae (Stainton, 1854) 101 genera, 1,866 species

Hepialidae (Stephens, 1829) 62 genera, 606 species

Immidae (Common, 1979) 6 genera, 245 species

Lacturidae (Heppner, 1995) 8 genera, 120 species

Lasiocampidae (Harris, 1841) 224 genera, 1,952 species

Lecithoceridae (Le Marchand, 1947) 100 genera, 1,200 species

Limacodidae (Duponchel, 1845) 301 genera, 1,672 species

Lyonetiidae (Stainton, 1854) 32 genera, 204 species

Megalopygidae (Herrich-Schäffer, 1855) 23 genera, 232 species

Micropterigidae (Herrich-Schäffer, 1855) 21 genera, 160 species

Mimallonidae (Burmeister, 1878) 27 genera, 194 species

Momphidae (Herrich-Schäffer, 1857) 6 genera, 115 species

Nepticulidae (Stainton, 1854) 13 genera, 819 species

Noctuidae (Latreille, 1809) 1,089 genera, 11,772 species

Nolidae (Bruand, 1847) 186 genera, 1,738 species

Notodontidae (Stephens, 1829) 704 genera, 3,800 species

Oecophoridae (Bruand, 1850) 313 genera, 3,308 species

Plutellidae (Guenée, 1845) 48 genera, 150 species

Prodoxidae (Riley, 1881) 9 genera, 98 species

Psychidae (Boisduval, 1829) 241 genera, 1,350 species

Pterolonchidae (Meyrick, 1918) 2 genera, 8 species

Pterophoridae (Latreille, 1802) 90 genera, 1,318 species

Pyralidae (Latreille, 1809) 1,055 genera, 5,921 species

Saturniidae (Boisduval, 1837) 169 genera, 2,349 species

Scythrididae (Rebel, 1901) 30 genera, 669 species

Sematuridae (Guenée,1858) 6 genera, 40 species

Sesiidae (Boisduval, 1828) 154 genera, 1,397 species

Sphingidae (Latreille, 1802) 206 genera, 1,463 species

Stathmopodidae (Janse, 1917) 44 genera, 408 species

Thyrididae (Herrich-Schäffer, 1846) 93 genera, 940 species

Tineidae (Latreille, 1810) 357 genera, 2,393 species

Tischeriidae (Spuler, 1898) 3 genera, 110 species

Tortricidae (Latreille, 1802) 1,071 genera, 10,387 species

Uraniidae (Leach, 1815) 90 genera, 686 species

Urodidae (Kyrki, 1988) 3 genera, 66 species

Xyloryctidae (Meyrick, 1890) 60 genera, 524 species

Yponomeutidae (Stephens, 1829) 95 genera, 363 species

Ypsolophidae (Guenée, 1845) 7 genera, 163 species

Zygaenidae (Latreille, 1809) 170 genera, 1,036 species

RESOURCES

BOOKS

Conner, W. E., ed. *Tiger Moths and Woolly Bears: Behavior, Ecology and Evolution of the Arctiidae* (Oxford University Press, 2009)

Crafer, T. *Foodplant List for the Caterpillars of Britain's Butterflies and Larger Moths* (Atropos Publishing, 2005)

Lees, D. C. and A. Zilli. *Moths: A Complete Guide to Biology and Behavior* (Smithsonian Books, 2019)

Marquis, R. J., S. C. Passoa, J. T. Lill, J. B. Whitfield, J. Le Corff, R. E. Forkner, and V. A. Passoa. *Illustrated Guide to the Immature Lepidoptera on Oaks in Missouri* (U.S. Department of Agriculture, Forest Service, Forest Health Assessment and Applied Sciences Team, 2019)

Miller, J. C. and P. C. Hammond. *Lepidoptera of the Pacific Northwest: Caterpillars and Adults* (USDA, 2003)

Miller, J. C., D. H. Janzen, and W. Hallwachs. *100 Caterpillars: Portraits from the Tropical Forests of Costa Rica* (Harvard University Press, 2006)

Porter, J. *The Colour Identification Guide to Caterpillars of the British Isles* (Brill, 2010)

Powell, J. A. and P. A. Opler. *Moths of Western North America* (University of California Press, 2009)

Scoble, M. J. *The Lepidoptera: Form, Function and Diversity* (Oxford University Press, 1992)

Wagner, D. L. *Caterpillars of Eastern North America: A Guide to Identification and Natural History* (Princeton University Press, 2005)

Wagner, D. L., D. F. Schweitzer, J. Bolling Sullivan, and R. C. Reardon. *Owlet Caterpillars of Eastern North America* (Princeton University Press, 2012)

Waring, P. and M. Townsend. *Field Guide to the Moths of Great Britain and Ireland.* 3rd edition (Bloomsbury, 2017)

SCIENTIFIC JOURNAL ARTICLES

Abe, T., M. Volf, M. Libra, R. Kumar, H. Abe, H. Fukushima, R. Lilip et al. "Effects of plant traits on caterpillar communities depend on host specialization." *Insect Conservation and Diversity* (2021)

Brown, J. W. "Patterns of Lepidoptera herbivory on conifers in the New World." *Journal of Asia-Pacific Biodiversity* 11: 1–10 (2018)

Greeney, H. F., L. A. Dyer, and A. M. Smilanich. "Feeding by lepidopteran larvae is dangerous: A review of caterpillars' chemical, physiological, morphological, and behavioral defenses against natural enemies." *Invertebrate Survival Journal* 9: 7–34 (2012)

Heppner, J. B. "Classification of Lepidoptera, Part 1. Introduction." *Holarctic Lepidoptera 5* (Suppl. 1), 148 pp. (1998)

Janzen, D. H. and W. Hallwachs. "To us insectometers, it is clear that insect decline in our Costa Rican tropics is real, so let's be kind to the survivors." *Proceedings of the National Academy of Sciences* 118 (2) (2021)

Kawahara, A. Y., D. Plotkin, M. Espeland, K. Meusemann, E. F. A. Toussaint, A. Donath, F. Gimnich et al. "Phylogenomics reveals the evolutionary timing and pattern of butterflies and moths." *Proceedings of the National Academy of Sciences* 116 (45): 22657–22663 (2019)

Pabis, K. "What is a moth doing under water? Ecology of aquatic and semi-aquatic Lepidoptera." *Knowledge & Management of Aquatic Ecosystems* 419:42 (2018)

Van Ash, M. and M. E. Visser. "Phenology of forest caterpillars and their host trees: The importance of synchrony." *Annual Review of Entomology* 52: 37–55 (2007)

ORGANIZATIONS DEDICATED TO THE STUDY AND CONSERVATION OF LEPIDOPTERA AND OTHER INSECTS

Amateur Entomologists' Society (UK)
https://www.amentsoc.org

Australian National Insect Collection
http://www.csiro.au/en/Research/Collections/ANIC

Buglife (UK)
https://www.buglife.org.uk

Les Lépidoptéristes de France
https://www.lepidofrance.com

The Lepidopterists' Society of Africa
http://www.lepsocafrica.org

The Lepidopterists' Society (USA)
https://www.lepsoc.org

McGuire Center for Lepidoptera & Biodiversity (USA)
https://www.floridamuseum.ufl.edu/mcguire

UKMoths
https://ukmoths.org.uk

Xerces Society (USA)
http://www.xerces.org

USEFUL WEBSITES

African Moths
https://www.africanmoths.com

Australian Caterpillars and their Butterflies and Moths
http://lepidoptera.butterflyhouse.com.au

BOLD/Barcode of Life Data Systems
https://www.boldsystems.org

BugGuide
http://www.bugguide.net

Butterflies and Moths of North America
http://www.butterfliesandmoths.org

Eggs, Larvae, Pupae and Adult Butterflies and Moths [UK]
http://www.ukleps.org/index.html

GBIF/Global Biodiversity Information Facility
https://www.gbif.org

HOSTS—a Database of the World's Lepidopteran Hostplants
http://www.nhm.ac.uk/our-science/data/hostplants

i-Naturalist
https://www.inaturalist.org

The Kirby Wolfe Saturniidae Collection
http://www.silkmoths.bizland.com/kirbywolfe.htm

Larvae of North-European Lepidoptera
http://www.kolumbus.fi/silvonen/lnel/species.htm

Lepidoptera and their ecology
http://www.pyrgus.de/index.php?lang=en

North American Moth Photographers Group
https://mothphotographersgroup.msstate.edu

Sphingidae of the Western Palearctic
http://tpittaway.tripod.com/sphinx/list.htm

Wikipedia Lepidoptera page
https://en.wikipedia.org/wiki/Lepidoptera

INDEX

E

F

G

H

I

J

K

L

ACKNOWLEDGMENTS

For me, this book stems from my lifelong enjoyment of Lepidoptera, which for my family—first my parents, and then my wife and children—has meant a correspondingly long history of me being distracted from "normal" life by this wonderful group of insects. Hence, I would like to gratefully acknowledge my family's continuous tolerance and love, which has allowed me to engage in my favorite pastimes of rearing, photographing, and studying moth caterpillars. I am grateful to John Heppner who encouraged me to undertake this writing project; to David Weiner, whose poignant question in a casual conversation is responsible for carnivorous plant-eating moths featuring in the book; and, to the late Tom Emmel, a friend, an advisor, and a colleague, for his companionship during many Lepidoptera research expeditions. Both Rachel Warren Chadd and I would like to thank Kate Shanahan, Caroline Earle, and the rest of the UniPress team for their encouragement throughout this project and the stunning design that has so helped bring the text to life.

Andrei Sourakov

PICTURE CREDITS

Illustrations on pages 17, 24T, 26, 38, 43, 44T, 46, 52T, 78, 82L, 83B, 99, 102, 104, 106, 108, 112, 114, 142, 144, 148, 150, 152, 164, 182, 184, 188, 212, 214, 242, 248, 250, 256, 267, 272, 274, 276 by John Woodcock.

The publisher would like to thank the following for permission to reproduce copyright material:

Adobe Stock /Aggi Schmid: 201; /Appreciate: 97TR; /creativenature.nl: 196–197B; /Golden Age Photos: 57T; /Ihorhvozdetskiy: 10; /Mary Evans Library: 93; /rostovdriver: 140; /Serhii: 41T; /Tim's insects: 197B; /willypd: 153.

Alamy Photo Library /Alex Fieldhouse: 41B; /Amazon-Images: 69; /Andrew Darrington: 221; /Andrew Newman Nature Pictures: 227; /Anna Seropiani: 266L; /Anton Sorokin: 68; /Avalon.red: 83R; /Bill Gorum: 234; /BIOSPHOTO: 230; /blickwinkel: 136L, 226, 243L; /Brian & Sophia Fuller: 176; /Buddy Mays: 59; /Buiten-Beeld: 251; /Chris Hellier: 151; /Custom Life Science Images: 76L, 76R; /Dan Gabriel Atanasie: 266R; /Danita Delimont: 247T; /Darren5907: 50; /David Whitaker: 9; /Denis Crawford: 129L, 155; /Dinodia Photos: 90; /DP Wildlife Invertebrates: 204T; /FLPA: 121B, 193, 208L, 249; /Frank Hecker: 20B; /George Grall: 166; /Grant Heilman Photography: 161B; /H Lansdown: 246; /imageBROKER: 119L, 134, 136–137T; imageBROKER/Reinhard Hölzl: 147; /ISI Foto: 175R; /Jeff Lepore: 138L, 141R, 187, 199R, 215, 240B; /John Cancalosi: 162; /Ken Barber: 236B; /Kike Calvo: 98; /Larry Corbett: 39; /Larry Doherty: 223; /Len Wilcox: 165; /Malcolm Schuyl: 60; /Marcos Veiga: 264; /Martin Williams: 233B; /mauritius images GmbH: 219; /Michelle Carden Photography: 236T; /Minden Pictures: 54, 95R, 95L, 172B, 205; /Nature Picture library: 55, 61, 88, 122T, 145, 149, 229B, 238; /Nigel Cattlin: 269; /Premaphotos: 125L; /Prisma by Dukas Presseagentur GmbH: 139T; /Rich Wagner: 171T; /Suzy Bennett: 169T; /Thomas Shjarback: 161T.

Andrei Sourakov 13 (col 2, T), 16, 18, 21TR, 21C, 22–23, 24B (both), 27T&B, 29, 30T, 32–33, 34–35, 36–37, 47T, 51R, 57B, 70B, 73T, 73B, 77, 79B, 82R, 97TL, 97CR, 163T; 197T, 200, 206, 217, 231BR, 241T&B, 244T T&B, 273.

Bart Wursten 126.

CSIRO /Entomology: 133T; /Geoff Clarke: 129R.

Dr. Alexandre Specht 233T.

Dreamstime /Artography: 167L; /Brett Hondow: 211B; /Cathy Keifer 19R, 52B; /Chan Yee Kee: 86; /Fritz Hiersche: 169B; /Henk Wallays: 229T; /Jason Ondreicka: 209R; /Javarman: 159 TR; /Joan Egert: 96; Manoj Kumar Tuteja: 2; /Matthew Omojola: 279; /Meisterphotos: 65; /Meoita: 208R; /Olha Pashkovska: 56; /Paul Reeves: 20T; /Rudolf Ernst: 89; /Rusty Dodson: 265; /Sandra Standbridge: 196T; /Sutisa Kangvansap: 32; /Verastuchelova: 135T.

Ervin Szombathelyi 255.

Flickr /Andy Reago & Chrissy McClarren: 231BL; /Brody J. Thomassen: 245R; /Clinton & Charles Robertson: 170; /Donald Hobern: 231TR, 231C; /janetgraham84: 232TL; /Katja Schulz: 275.

Iain Leach 253, 257.

Igor Siwanowicz 21B, 25, 28, 64, 143, 198, 210, 211T, 213.

iStock Dan Gabriel Atanasie: 268T; /ErikaMitchell: 199L; /LucynaKoch: 124; /MsScarlett: 167R.

John Horstman 12, 13 (col 2, B), 58, 66T, 66B, 70T, 75, 79T, 91, 92, 103, 105, 107, 109, 131.

Kirby Wolfe 71L, 101, 115, 172T.

Lary Reeves 13 (col 3, C), 40, 44, 47B, 72.

Nature in Stock /Gianpiero Ferrari/FLPA: 119R, 239B; /Hugh Lansdown /FLPA: 30B; /Patricio Robles Gil/Sierra Ma/Minden Pictures: 113; /Piotr Naskrecki/Minden Pictures: 111; /Tom Kruissink: 202.

Nature Picture Library /Hans Christoph Kappel: 42TR; /Jose B. Ruiz: 189; /Jussi Murtosaari: 177; /Nick Upton: 53; /Robert Thompson: 67, 173, 183.

Sergei Drovetski 235.

Shutterstock /aDam Wildlife: 62; /Andrei Stepanov: 178T; /Arran Jesson: 237L; /Bernard Barroso: 228T; /Bridget Calip: 118; /coffee prince: 3; /COULANGES: 209L; /damann: 231T; /Dark Egg: 240T; /Dean Pennala: 195; /DJTaylor: 243R; /Erik Karits: 11; /Filip Fuxa: 4C, 13 (col 4, T); /Gucio_55: 237R; /idiz: 133B; /Independent birds: 4T; /Ireneusz Waledzik: 42TL; /irisphoto1: 125R; /J Need: 13 (col 3, T); /Jay Ondreicka: 231CL; /Karel Gallas: 160; /Kevin Wells Photography: 94; /Lavandina: 228B; /Marcos Cesar Campis: 268B; /Marek R. Swadzba: 135B, 277; /Margus Vilbas Photography: 174; /Matee Nuserm: 4BR; /Matt Jeppson: 207 (all); /Melinda Fawver 4TC, 13 (col 1, C), 13 (col 1, B), 178B; /Mikhail Melnikov: 4BL, 13 (col 2, C); /Mirek Kijewski: 194; /MRS. NUCH SRIBUANOY: 259; /NERYXCOM: 13 (col 4, B); /Nova Patch: 245L; /Paul J Hartley: 192L; /pixelworlds: 81; /Poppap pongsakorn: 138R; /Protasov AN: 4BC, 13 (col 3, B), 168, 239T; /Rainer Lesniewski: 127R; /Rajavel P: 171B; /Randy Bjorklund: 179; /Raphael Comber Sales: 122B; /romm: 4B; /Sandra Standbridge: 204B; /Sarah2: 137R; /ScottYellox: 130; /ShaunWilkinson: 4TR; /Tawansak: 63; /Tomasz Klejdysz: 13 (col 4, C), 270, 271L&R; /totajla: 128–129B; /Vadim Nefedoff: 158L; /Vinicius Bacarin: 120–121; /Vitalii Hulai; 8, 13 (col 1, T); /Wildnerdpix: 141L; /William Cushman: 132; /xpixel: 159C; /Zety Akhzar: 203.

Thomas Shahan 51L.

Tom Astle 45, 71R, 87, 159B, 175L, 181, 185, 261.

USDA Agricultural Research Service /Peggy Greb: 163B.

Wikimedia Commons /Charles J. Sharp: 123, 247B; /Harald Süpfle 19L; Rotational: 127L.